THE POINT OF LAW
HEALTH AND SAFETY
AT WORK etc.
ACT 1974

explained by **BRADLEY** LLB(Hons) DMA PGCE MIOSH Solicitor

London: The Stationery Office

Applications for reproduction should be made in writing to The Stationery Office Limited, St Crispins, Duke Street, Norwich NR3 1PD.

The information contained in this publication is believed to be correct at the time of manufacture. Whilst care has been taken to ensure that the information is accurate, the publisher or author can accept no responsibility for any errors or omissions or for changes to the details given.

A CIP catalogue record for this book is available from the British Library
A Library of Congress CIP catalogue record has been applied for

ISBN 0 11 702810 X

This book is dedicated by Ken Bradley with all my love to MARGARET JANE MARTIN (nee AURELIUS)

Published by The Stationery Office
and available from:

The Stationery Office
(mail, telephone and fax orders only)
PO Box 29, Norwich, NR3 1GN
Telephone orders/General enquiries 0870 600 5522
Fax orders 0870 600 5533
www.thestationeryoffice.com

The Stationery Office Bookshops
123 Kingsway, London WC2B 6PQ
020 7242 6393 Fax 020 7242 6394
68-69 Bull Street, Birmingham B4 6AD
0121 236 9696 Fax 0121 236 9699
33 Wine Street, Bristol BS1 2BQ
0117 926 4306 Fax 0117 929 4515
9-21 Princess Street, Manchester M60 8AS
0161 834 7201 Fax 0161 833 0634
16 Arthur Street, Belfast BT1 4GD
028 9023 8451 Fax 028 9023 5401
The Stationery Office Oriel Bookshop
18-19 High Street, Cardiff CF1 2BZ
029 2039 5548 Fax 029 2038 4347
71 Lothian Road, Edinburgh EH3 9AZ
0870 606 5566 Fax 0870 606 5588

The Stationery Office's Accredited Agents
(see Yellow Pages)

and through good booksellers

Printed in the United Kingdom by the Stationery Office Limited
TJ005616   C10   11/01   19585   662314

Contents

The Health and Safety at Work etc Act 1974, Explained

The Guide to the Health & safety at Work etc. Act 1974 and the associated Management of Health & Safety at Work Regulations, 1999 was written by Ken Bradley, LLB(Hons) DMA PGCE MIOSH Solicitor of 'Winstanley' 1, Tusting Close, Sprowston, Norwich, Norfolk, NR7 8TD – telephone (01603) 406785.

Ken studied for and gained a law degree at Birmingham University (1969-1972) and went on to gain his Solicitor's qualification via the College of Law, Guildford. After practising in local government he entered Further and Higher Education as a lecturer. For 24 years he taught a wide range of law, health and safety, management and trade union courses at all levels to over 5000 students.

Ken came to specialise in health and safety law, employment law and consumer law and has practical as well as academic experience.

He has written books and articles on law, health & safety and history and has been an examiner for NEBOSH and AQA.

He is currently a tutor for RRC Business Training Ltd and is a Health and Safety Consultant for Suffolk County Council and on his own account. He also has other small business interests and is a legal advisor also.

He has always tried to communicate the law to others in a way that is intelligible and in recognition that most people are not lawyers.

Disclaimer

This publication is intended to provide a brief commentary on the Health and Safety at Work etc. Act 1974 and the Management of Health and Safety at Work Regulations 1999 and should not be relied upon by anyone without taking further legal advice.

NB. The Act is believed to be up to date (including all repeals etc) up to *October 31st 2001*. Caution should be exercised throughout as Parliament may make further repeals as time goes by.

Introduction

The Health & Safety at Work etc. Act, 1974(HSW Act) applies to most workplaces in England, Wales and Scotland . The Act does not apply to Northern Ireland but broadly similar provisions apply there. The Republic of Ireland has broadly similar laws also.

A feature of the Act is that, although Employers and the Self-employed have the main duties, others have duties also. These include, contractors, landlords, manufacturers, suppliers, designers, importers, exporters and employees. Managers and supervisors may have individual liability also.

The Act enacted, with all Party support, the recommendations of The Robens Report(1972, Cmnd 5034). The Royal Commission, chaired by Lord Robens, was established in June 1970 by Barbara Castle, the then Secretary of State for Employment and Productivity. The Commission had wide terms of reference to inquire into the state of legislation on health, safety and welfare and to make recommendations for reform.

Existing legislation was found to be faulty in several respects including:
Its piecemeal and reactive nature over two centuries
It's non application to approximately 20% of workplaces including hospitals, schools, colleges, laboratories, etc.
Factories, Offices, Shops, Railway premises, Mines and Quarries and agriculture had specialist legislation but many industries were covered by no safety legislation at all
It's complicated nature and applicability to particular machines, processes, individuals
A confusing set of enforcement authorities

The HSW Act seeks to overcome these anomalies and takes the line that everyone is involved in health safety and welfare to a greater or lesser extent and should share responsibilities and liabilities. All of industry is covered, although there are some small exceptions.
The Act lays down broad organisational requirements on employers and the self employed especially and insists on forward planning and a risk assessment approach before accidents happen in the workplace.
The duties are backed up by a range of criminal sanctions largely enforced by the Health & Safety Executive (a new body created by the Act) by various notices but ultimately via the Magistrate's' Courts or (less often) the Crown Court.

Since the UK joined the European Union (EU) in 1972, health and safety legislation has been influenced heavily by EU law and many of the Regulations which supplement the Act and fill in the details, derive from EU Directives.

In this book the focus will be on the Act and the associated MHSWR. It must be borne in mind, however, that other laws apply in the health, safety and welfare field including;

The Common Law:
(1)   Criminal Law – manslaughter especially

(2) Civil Law – law of contract, law of torts/civil wrongs (i.e. negligence, nuisance, trespass, etc.) and breach of statutory duty. Health & safety law also overlaps with other laws such as those on employment and agency.

Other Statutes

(1) Criminal law – eg Environment Protection legislation

(2) Civil Law – Occupiers Liability Acts 1957 and 1984

Other Regulations/Statutory Instruments

There are many regulations made under the HSW Act applying to particular hazards and/or procedures in the workplace.
These include;
The Classification of Hazards, Information and Packaging Regulations,1999(CHIP)
The Confined Spaces Regulations,1997
The Control of Substances Hazardous to Health Regulations,1999(COSHH)
The Electricity at Work Regulations,1989(EAWR)
The Health & Safety (Display Screen Equipment) Regulatons,1992(DSER)
The Lifting Operation and Lifting Equipment Regulations,1998(LOLER)
The Management of Health and Safety at Work Regulations,1999(MHSWR)
The Manual Handling Regulations,1992(MHR)
The Noise at Work Regulations,1989(NAWR)
The Personal Protective Equipment Regulations,1992(PPER)
The Provision and Use of Workplace Equipment Regulations, 1998(PUWER)
The Working Time Regulations 1998(as amended)(WTR)
The Workplace (Health Safety and Welfare) Regulations 1992(WPHSWR).

It is breach of the criminal law to infringe these regulations. A breach of statutory duty action under the regulations may be brought in some instances by victims suing in the civil courts.

European Union Law

Most EU laws are enacted into UK law. If there are omissions, however, EU law prevails over internal law and challenges can be made in internal courts and ultimately to the European Court of Justice.

The focus in this book will be on the HSW Act and the MHSWR which provide criminal sanctions for those in breach. The other laws referred to above are beyond the scope of this book but it is important that there is an awareness of these other laws and that other publications and appropriate legal advice be consulted.

The Act has been amended from time to time since 1974 but most of its provisions remain intact. Important changes over the years have included:
Bringing young people on training schemes under the protection of the Act
Inclusion of offshore oil and gas installations
The bringing together of most inspectorates under the Health & safety Executive

Changes as a result of EU law, many of which are enacted into new Regulations under the HSW Act rather than by an amendment of the Act itself.

Despite all Party support for the Act in 1974 various attempts have been made at de-regulation. However, the supremacy of EU Law has limited these aims somewhat.

Twenty five years later, it has to be said that the Act has not prevented some major disasters and some of the enforcement provisions leave a lot to be desired. There are still too few HSE inspectors on the ground to enforce the Act and the associated regulations.

Many employers and others are taking their legal responsibilities seriously, however, and no doubt much good has been done by the passing of this legislation. The law is being strengthened all the time. At the time of writing there are proposals for a new criminal offence of 'corporate killing' or 'corporate manslaughter' which if enacted and used will overcome the present problems whereby only individuals can be prosecuted for manslaughter.

Despite the flaws, one can never be certain when an HSE inspector may call and use his or her enforcement powers and ultimately criminal prosecution in the criminal courts. In extreme cases this could lead to the closure of one's business. For these and many other reasons compliance is essential.

Health and Safety at Work etc Act 1974

1974 CHAPTER 37

An Act to make further provision for securing the health, safety and welfare of persons at work, for protecting others against risks to health or safety in connection with the activities of persons at work, for controlling the keeping and use and preventing the unlawful acquisition, possession and use of dangerous substances, and for controlling certain emissions into the atmosphere; to make further provision with respect to the employment medical advisory service; to amend the law relating to building regulations, and the Building (Scotland) Act 1959; and for connected purposes   [31st July 1974]

BE IT ENACTED by the Queen's most Excellent Majesty, by and with the advice and consent of the Lords Spiritual and Temporal, and Commons, in this present Parliament assembled, and by the authority of the same, as follows:-

PART I

HEALTH, SAFETY AND WELFARE IN CONNECTION WITH WORK, AND CONTROL OF DANGEROUS SUBSTANCES AND CERTAIN EMISSIONS INTO THE ATMOSPHERE

Section 1

The purpose of this section is to make it clear that the overriding purpose of the Act is to protect the health, safety and welfare of 'persons at work'.

'Persons at work' is not defined in the Act but would include: Employers, the self-employed, employees, contractors, visitors, etc. etc. .Anyone in the workplace at any time is a 'person at work.'

'Health' includes protection from accidents, industrial diseases, exposure to dangerous chemicals and substances and other agents (e.g. noise), protection from psychological harm (e.g. stress), musculoskeletal disorders etc. etc.

'Safety' means taking the appropriate standard of care to protect persons at work from death and accidents in the workplace

'Welfare' includes such things as sanitary conveniences, washing facilities, drinking water, accommodation for clothing, changing facilities, facilities for rest and to eat meals and special provisions to protect young persons and pregnant women and nursing mothers at work.

'Work' is defined in section 52(1)(a) as: '..... work as an employee or as a self employed person'

However, by s52(1)(a) 'an employee is at work throughout the time when he is in the course of his employment, but not otherwise. Employers' are not therefore liable for actions when an employee is not actually employed on the job he is employed to do and nor where the employee

is on break time (provided it is his/her own time) or where the employee has gone off on 'a frolic of his own' during working hours without permission. There is much case law on this in the law of torts (civil wrongs) which the judges would probably apply here also even though the HSW Act is a criminal law piece of legislation. There is also specific case law on the Act defining 'course of employment') see Coult v Szuba(1982)ICR 380)

Police constables are 'at work throughout the time they are on duty but not otherwise (see s52(1(bb)).

'self employed persons' are 'at work throughout such time as he devotes to work as a self employed person.' (see s52(1)(c)). This might extend to doing the accounts in the evenings.

'Persons other than persons at work' (see s1(1)(b) refers to the general public including passers by who may be affected by activities at work (e.g. pollution, explosions, etc.).

This section makes it clear that everyone is protected to some extent by the provisions of the Act.

'Employees' are defined by s53(1) as '..an individual who works under a contract of employment (or is treated by s51A) as being an employee, and related expressions shall be construed accordingly. Again, there is much case law on the definition of an employee in cases of doubt.

'Work' is further defined in certain regulations and case law For example, by the Health & Safety (Training for Employment) Regulations 1990 'work' includes relevant training within the meaning of the regulations. 'Work' also includes 'any activity involving genetic manipulation' and the 'keeping or handling of a listed pathogen'. Regulations applying to Offshore installations, first aid and Control of Substances Hazardous to Health also define 'work'.

'Self employed person' is defined in s53(1) as '.. an individual who works for gain or reward otherwise than under a contract of employment, whether or not he himself employs others'

This obviously includes a lot of small businesses including the 'one man band.'

Preliminary

1   Preliminary

(1)   The provisions of this Part shall have effect with a view to–

    (a) securing the health, safety and welfare of persons at work;

    (b) protecting persons other than persons at work against risks to health or safety arising out of or in connection with the activities of persons at work;

    (c) controlling the keeping and use of explosive or highly flammable or otherwise dangerous substances, and generally preventing the unlawful acquisition, possession and use of such substances; ...

S1(1)(c) is fairly self explanatory but related legislation has provided some detailed rules now which must be taken into account by those affected. This legislation includes:
The Explosives legislation
The Firearms legislation
The Highly Flammable Liquids and Liquefied Petroleum Gas Regulations, 1972
The Control of Substances Hazardous to Health Regulations, 1999.

(d) ...

S1(1)(d) now repealed but its provisions as to 'emission into the atmosphere of noxious or offensive substances' is now substantially re-enacted in The Environmental Protection Act, 1990 which overlaps considerably with HSW Act.

(2) The provisions of this Part relating to the making of health and safety regulations... and the preparation and approval of codes of practice shall in particular have effect with a view to enabling the enactments specified in the third column of Schedule 1 and the regulations, orders and other instruments in force under those enactments to be progressively replaced by a system of regulations and approved codes of practice operating in combination with the other provisions of this Part and designed to maintain or improve the standards of health, safety and welfare established by or under those enactments.

S1(2) is designed to ensure that older Health and Safety legislation (e.g. Factories Act, 1961 etc.) is progressively replaced by regulations made under s15 of this Act thus bringing everything under the HSW Act 'umbrella' and applying to all workplaces generally. This process is now (2001) largely complete and most of the older legislation has been repealed.

(3) For the purposes of this Part risks arising out of or in connection with the activities of persons at work shall be treated as including risks attributable to the manner of conducting an undertaking, the plant or substances used for the purposes of an undertaking and the condition of premises so used or any part of them.

S1(3) This section makes it clear that person other than persons at work are also protected from risks emanating from:
The manner of conducting an undertaking
Any plant or substances used in the undertaking
The conditions of the premises.
Thus, employers' and other have to take these into account also.

(4)   References in this Part to the general purposes of this Part are references to the purposes mentioned in subsection (1) above.

> *S1(4) wherever in Part 1 of the Act (sections 1 to 54) there is reference to the words 'the general purposes of this Part', the 'purpose' shall include those spelled out in section 1(1)(a) to (c) and nothing else.*

GENERAL DUTIES

2   General duties of employers to their employees

> *This is one of the most important sections of the Act and lays down the general duties of all Employers to all employees. More specific and detailed duties are contained in the various Regulations made under s15 of the Act.*
> *'Health' 'Safety' and 'welfare' are defined in the notes to section 1 above.*

(1)   It shall be the duty of every employer to ensure, so far as is reasonably practicable, the health, safety and welfare at work of all his employees.

> **S2(1)**
> *Under this sub section there is an absolute duty on all employers to take steps to protect the health safety and welfare of all employees. However, the actual steps that an employer must take are qualified by the words 'so far as is reasonably practicable'. The Act does not define these words but various decided cases have supplied this omission as follows:*
>
> *In Edwards v National Coal Board (1949) 1 All ER 743 CA, Lord Justice Asquith stated (at p747):*
> *'Reasonably practicable, as traditionally interpreted, is a narrower term than 'physically possible' and implies that a computation must be made in which the quantum of risk is placed in one scale and the sacrifice, whether in money time or trouble, involved in the measures necessary to avert the risk is place in the other; and that, if it be shown that there is a gross disproportion between them, the risk being insignificant in relation to the sacrifice, the person upon whom the duty is laid discharges the burden of proving that compliance was not reasonably practicable. This computation falls to be made at a point of time anterior to the happening of the incident complained of.'*
> *This statement has been approved in several other cases including the House of Lords in Marshall v Gotham Co Ltd (1954) 1 All ER 937 HL.*
>
> *In short, an employer must do what any reasonable employer would do to protect the employee in question bearing in mind the nature of the hazard, persons at risk, the degree of danger, etc. measured against the cost and practicability of safety measures/control measures. This is reinforced by Regulations made under the Act which require risk assessments.*

Employers as a basic minimum must follow the law, any relevant Code of Practice and Guidance notes (or be doing something at least as good) and following best practice and advice).

(2)   Without prejudice to the generality of an employer's duty under the preceding subsection, the matters to which that duty extends include in particular–

S2(2)

This sub section spells out a number of situations which the Employer must particularly pay regard to. This does NOT take away from the comprehensive overall duty in s2(1) and many other matters will be encompassed by s2(1) than those spelled out in s2(2)(a) to (e), supplemented by the many Regulations made under s15.

(a) the provision and maintenance of plant and systems of work that are, so far as is reasonably practicable, safe and without risks to health;

S2(2)(a)

This subsection makes it clear that plant provided must be maintained properly and regularly and that systems of work should be devised that provide for employee safety and without risks to health so far as is reasonably practicable.
Contractor's employees should be included in 'safe systems' also (see R v Swan Hunter Shipbuilders Ltd (1982) 1 All ER 264, CA.
'Plant' is defined in s53(1) as including 'any machinery, equipment or appliance.'
This would include plant wherever situated including offices, workshops, factories, laboratories, schools, colleges, hospitals, etc. etc.
More detailed provisions concerning 'work equipment', 'lifts' and 'lifting operations' is contained (inter alia) in the following Regulations which must also be implemented by employers:
The Provision and Use of Work Equipment Regulations, 1998 (PUWER)
The Lifts Regulations, 1997
The Lifting Operations and Lifting Equipment Regulations 1998 (LOLER).

'Systems of work' is not defined by the Act but this section requires that the employer devises a safe system of working for every piece of plant, process etc. especially where employees could be injured or killed through the lack of such a system. These systems should be written and should take account of and follow on from the Health and Safety Policy (see S2(3) below) and appropriate risk assessments made under the various Regulations made under s15. All of these documents should be filed carefully and be available to all who need to consult them.
Where necessary 'permit to work' systems should be devised for the more dangerous processes (e.g. Chemicals) and plant (e.g. Fork lifts, guillotines, etc.) whereby workers are trained to an appropriate standard by a 'competent person' who then awards them a 'permit to work'.

This permit should be shown to the employee's supervisor before being allowed to work in the particular area or on the particular piece of plant. All dangerous equipment and processes should be 'locked off' so that unauthorised persons with a 'permit to work' cannot gain access.

'So far as is reasonably practicable' – see notes to s2(1) above.

There have been many prosecutions under this sub section.

(b) arrangements for ensuring, so far as is reasonably practicable, safety and absence of risks to health in connection with the use, handling, storage and transport of articles and substances;

S2(2)(b)

This sub section applies to all articles and substances at work.

'Article for use at work' is defined in s53(1) as:

'(a) any plant designed for use or operation (whether exclusively or not) by persons at work, and

(b) any article designed for use as a component in any such plant.'

Thus, any machinery, equipment or appliance is an article for use at work whether or not employees actually work it some of the time and, for example, members of the public at other times (e.g. bank cash machines etc.).

The PUWER Regulations 1998 would also apply to 'articles at work'.

'Substance' is defined in s53(1) as, 'any natural or artificial substance, whether in solid or liquid form or in the form of a gas or vapour'. Natural substances would include water and asbestos. Artificial substances would include all paints, glues, resins etc.

'Substance for use at work' is defined in s53(1) as 'any substance intended for use (whether exclusively or not) by persons at work'.

Substances are further defined and controlled by (inter alia):

The Classification of Hazards, Information and Packaging Regulatons, 1999 (CHIP)

The Control of Substances Hazardous to Health Regulations 1999(COSHH)

The Control of Asbestos at Work Regulations, 1987(as amended 1992)

The Control of Lead at Work Regulations, 1998

The Ionising Radiation Regulations, 1985

The subsection makes it clear that in all operations involving articles and substances including the use, handling, storage and transport safe practices should be devised so far as is reasonably practicable to protect employees at work.

'Articles' are further controlled by, for example, the PUWER Regulations 1998 and additionally there are several regulations governing the transport of articles and substances.

(c) the provision of such information, instruction, training and supervision as is necessary to ensure, so far as is reasonably practicable, the health and safety at work of his employees;

S2(2)(c)

This very important sub section makes it clear that for all activities at work (but especially the more dangerous ones) employers must ensure so far as is reasonably practicable that employees are properly supervised by competent supervisors, and are given appropriate information, instruction and training so as to ensure their health and safety at work. Thus, new employees should be given induction training which should be updated regularly. All employees should receive appropriate training on all articles and substances and systems they have to use at work and how to operate them safely. This too should be updated at regular intervals, especially if new information comes out.

Contractor's employees should also be given the benefit of this sub section (see R v Swan Hunter Shipbuilders Ltd(1982) 1 All ER 264,CA.

Most regulations made under s15 repeat this requirement in more detail. Training records should be kept for each employee. There is a close link with risk assessment here also. Many prosecutions have been brought under this sub section against employers who have not obeyed the law.

(d) so far as is reasonably practicable as regards any place of work under the employer's control, the maintenance of it in a condition that is safe and without risks to health and the provision and maintenance of means of access to and egress from it that are safe and without such risks;

S2(2)(d)

Under this subsection the workplace itself (provided it is under the Employers control) must be maintained in a safe condition. This means that the structure and fixtures particularly must be maintained so as not to collapse and cause injury or death.

Maintenance must also encompass cleaning etc. to ensure that diseases are not caused (e.g. Legionella, hepatitis B, Weils disease, asthma & other lung diseases, etc. etc.).

Further requirements on maintenance are contained in (inter alia);
The Workplace (Health Safety and Welfare) Regulations,1992
The COSHH Regulations,1999

The access to and exits from the workplace must also be maintained in a safe state. This latter requirement may overlap with requirements under the Fire Prevention (Workplace) Regulations 1997 and other fire safety legislation.
It would also encompass ensuring that traffic is not a hazard to pedestrians.

(e) the provision and maintenance of a working environment for his employees that is, so far as is reasonably practicable, safe, without risks to health, and adequate as regards facilities and arrangements for their welfare at work.

S2(2)(e)

Under this sub section the whole working environment must be safe for employees and free of anything which may cause injury, disease or death. This environment must also be maintained which involves cleaning processes and maintenance of such things as ventilation etc.

Welfare facilities must be provided and maintained.

These are defined above (see notes to s1) and further detailed requirements are contained in (inter alia);

The Workplace (Health Safety and Welfare) Regulations, 1992

The COSHH Regulations 1999.

(3) Except in such cases as may be prescribed, it shall be the duty of every employer to prepare and as often as may be appropriate revise a written statement of his general policy with respect to the health and safety at work of his employees and the organisation and arrangements for the time being in force for carrying out that policy, and to bring the statement and any revision of it to the notice of all his employees.

S2(3)

Under this sub section there is an absolute duty on all employers to prepare a written statement of his Health & Safety Policy covering:

The health and safety at work of his employees

The safety organisation within the Company, firm or business

The health & safety arrangements in the workplace for carrying out the policy.

To review it on a regular basis.

The Management of Health & Safety at Work Regulations, 1999 spell these requirements out in more detail (see pages 134–170)

This statement and any revision of it must be brought to the attention of all employees. This is usually done in a joiners pack and reinforced in an Induction training session on health & safety.

Small businesses may get a draft policy and guidance from HSE Books and fill in the gaps.

By the Employers' Health & safety Policy Statements (Exception) Regulations, 1975 (SI 1975/1584) employers who employ less than five employees in an undertaking at any one time are exempt from the requirement to record the policy in writing. This does NOT exempt small employers from assessing risks to employees and others and in cases of doubt it is best to record everything in writing particularly where the HSE provides useful forms and guides.

NB. 'Absolute duties' are those that the employer must implement and for which there is little or no defence for non compliance. The words, 'shall', 'must', 'will' etc. signifies that there is an absolute duty under the Act. This should be contrasted with the other 'qualified duties' e.g. 'so far as is reasonably practicable', 'best practicable means', 'so far as is reasonable' etc.

(4)   Regulations made by the Secretary of State may provide for the appointment in prescribed cases by recognised trade unions (within the meaning of the regulations) of safety representatives from amongst the employees, and those representatives shall represent the employees in consultations with the employers under subsection (6) below and shall have such other functions as may be prescribed.

S2(4)

This sub section allowed to Secretary of State (for Education & Employment) to make Regulations allowing for the election of trade union safety representatives by recognised trade unions to represent employees in the workplace and have various functions laid down in the Regulations.

The regulations covering these matters are: The Safety Representatives and Safety Committees Regulations, 1977 (as amended) (SRSC).

Broadly, if an employer recognises a trade union for negotiating purposes on pay and conditions in the workplace, the union has the right to elect safety representatives with statutory functions (including time off with pay to do the job and to attend union training courses). Those safety representatives must be consulted on most things to do with health & safety in the workplace.

Statistics have proved that where these union safety representatives are in place and active, the workplace tends to be a safer workplace as a result.

'Recognition' may now be gained by trade unions using the law as laid down in the Employee Relations Act, 1999. Note that only 'recognised' unions can claim these rights.

Where no safety representatives exist, the employer must still consult the workforce on health and safety issues.

(5)   ...

S2(5)

This subsection (now repealed?) is in substantially similar terms to s2(4) but allows for the appointment of non union representatives.

For non unionised workplaces or workplaces where there is a union but it is not 'recognised' the Health & safety (Consultation with employees) Regulations, 1996 (HASCER) allow for the appointment of 'Representatives of Employee Safety'. These representatives have similar (but slightly diluted) rights to those afforded to union safety reps under SRSC.

(6)   It shall be the duty of every employer to consult any such representatives with a view to the making and maintenance of arrangements which will enable him and his employees to co-operate effectively in promoting and developing measures to ensure the health and safety at work of the employees, and in checking the effectiveness of such measures.

S2(6)

There is an absolute duty on all employers to consult with:

Recognised union safety representatives

Representatives of employee safety

The whole workforce where there are no representatives,

with a view to the making and maintenance of health and safety arrangements which will enable the employers and employees to co-operate effectively to promote and develop measures to ensure the health and safety at work of all the employees and to check the effectiveness of such measures.

The Regulations mentioned above (see s2(4) and 2(5) contain detailed provisions).

This reinforces the view that health and safety must be an area in which employers and employees co-operate to mutual advantage; it should not be a 'battle ground.'

NB. 'Absolute duties' are defined in the note to s2(3) above

(7) In such cases as may be prescribed it shall be the duty of every employer, if requested to do so by the safety representatives mentioned in [subsection (4)] above, to establish, in accordance with regulations made by the Secretary of State, a safety committee having the function of keeping under review the measures taken to ensure the health and safety at work of his employees and such other functions as may be prescribed.

S2(7)

This subsection allows for union safety representatives to request of the employer that a safety committee be established for the particular organisation.

SRSC regulations (see reg. 9) allow for 2 safety representatives to request in writing such a Committee. On receipt, the employer must consult as to the constitution, functions and membership of the Safety Committee and must in any case establish it within 3 months of the request being made. A Code of Practice/Guidance Notes under SRSC suggests membership and functions.

HASCER does not allow for employee representatives to request a Safety Committee.

3 General duties of employers and self-employed to persons other than their employees

(1) It shall be the duty of every employer to conduct his undertaking in such a way as to ensure, so far as is reasonably practicable, that persons not in his employment who may be affected thereby are not thereby exposed to risks to their health or safety.

S3(1)–EMPLOYERS DUTIES TO OTHERS

This sub section places a duty on Employers to conduct their businesses with proper regard to health and safety to other people/non employees who could be affected by work activities so far as is reasonable practicable.

These other people include: contractors and their employees, visitors, delivery persons, neighbours, the general public, etc. etc.

This is reinforced by the risk assessment provisions of most regulations whereby risk assessments must also take account of such persons as well as employees under s2.

Definitions: 'employee' see s53(1) and the note to s2 above

'so far as is reasonably practicable' – see note to s2 above.

'Risks' – see note to s3(2) below.

(2) It shall be the duty of every self-employed person to conduct his undertaking in such a way as to ensure, so far as is reasonably practicable, that he and other persons (not being his employees) who may be affected thereby are not thereby exposed to risks to their health or safety.

S3(2)–DUTIES OF SELF-EMPLOYED PERSONS

This subsection applies to self-employed persons who must so far as is reasonably practicable take steps to ensure that the following people are not exposed to risks to their health and safety:

The self-employed person himself or herself

Other people/non employees (as in s 3(1)

Risk assessments should similarly cover such people.

NB. The self-employed person's own employees should be covered by S2. Note also that even if the only injury is to the self-employed person himself/herself the HSE could still take action.

Definitions:

'Self-employed person' by s53(1) means: 'an individual who works for gain or reward otherwise than under a contract of employment, whether or not he himself employs others.'

This would include sole traders and partners in a partnership firm and appropriate people in unincorporated associations but NOT company directors etc. who are 'employees'.

See also s52(1).

Companies are NOT included in s3(2) as they are not 'individuals' and are included in s2 and s3(1) as 'Employers'.

Charities are probably excluded from this section especially if they do not set out to work for gain or reward but they may be 'Employers' covered by the duties in s2 and s3(1).

'So far as is reasonably practicable' – see note to s2 above.

'Risk' is widely interpreted in R v Board of Trustees of the Science Museum (1993) 3 All ER 853 CA, to encompass the mere possibility of danger. The HSE or CPS does not have to show there was an actual danger.

(3) In such cases as may be prescribed, it shall be the duty of every employer and every self-employed person, in the prescribed circumstances and in the prescribed manner, to give to persons (not being his employees) who may be affected by the way in which he conducts his undertaking the prescribed information about such aspects of the way in which he conducts his undertaking as might affect their health or safety.

> **S3(3)**
> *This subsection allows the Secretary of State (for Education and Employment) to make Regulations imposing duties on both employers and self-employed persons to give to persons not in their employment prescribed information about the way in which they conduct their undertaking which might affect their health and safety.*
> *No regulations have been made to date under this section although some other regulations do envisage the visitors etc. so far as is reasonably practicable being provided with some information (e.g., MHSWR, COSHH, etc)*

4 General duties of persons concerned with premises to persons other than their employees

(1) This section has effect for imposing on persons duties in relation to those who–

   (a) are not their employees; but

   (b) use non-domestic premises made available to them as a place of work or as a place where they may use plant or substances provided for their use there,

   and applies to premises so made available and other non-domestic premises used in connection with them.

> **S4(1)**
> *This sub section makes it clear that s4 applies broadly to landlords and similar persons who have duties towards those who are not employed by them but who lease or rent or hire non domestic premises as a place of work or as a place where they may use plant or substances on the site. The lessee etc. would generally be a person conducting a business from those leased premises. That person has duties under s2 & s3(1). Additionally, however, the landlord may have some liability under s4 if the lease etc. provides for his maintaining the structure of the premises etc.*

(2) It shall be the duty of each person who has, to any extent, control of premises to which this section applies or of the means of access thereto or egress there from or of any plant or substance in such premises to take such measures as it is reasonable for a person in his position to take to ensure, so far as is reasonably practicable, that the premises, all means of access thereto or egress therefrom available for use by persons using the premises, and any plant or substance in the premises or, as the case may be, provided for use there, is or are safe and without risks to health.

S4(2)

This sub section imposes a duty on landlords etc. who have :

control of premises or any part of the premises

control of means of access to or egress from the premises

control of any plant or substance in the premises

to take such measures as it is reasonable for such a person to take to ensure so far as is reasonably practicable that the premises and any part of it, means of access and egress and any plant and substances on site are are safe and without risks to health to any person using the premises.

The courts will certainly look at provisions in leases and contracts as to landlords repairing obligations and in the absence of such documentation will look at usual landlord liability.

Definitions:

'Premises', 'non domestic premises', 'plant' and 'substance' are defined in s53(1).

'reasonable' – the courts will measures what is reasonable by what other reasonable landlords would do in similar circumstances.

'so far as is reasonably practicable' – see notes on s2 above.

'risk' – see note to s3(2) above.

'work' is defined in s52(1)

'persons' include legal persons (i.e. companies etc.) and natural persons (inc. sole traders, partnerships and unincorporated associations) – Interpretation Act,1978 s5 & Sched 1.

(3) Where a person has, by virtue of any contract or tenancy, an obligation of any extent in relation to–

   (a) the maintenance or repair of any premises to which this section applies or any means of access thereto or egress therefrom; or

   (b) the safety of or the absence of risks to health arising from plant or substances in any such premises;

   that person shall be treated, for the purposes of subsection (2) above, as being a person who has control of the matters to which his obligation extends.

S4(3)

This subsection makes it clear that even if a person is not a landlord but nevertheless has some obligations under a contract or tenancy to:

maintain the premises or any access or egress therefrom

repair the premises or any access or egress therefrom

he/she shall be treated as a 'landlord' under s4(2) and will have the obligations and responsibilities set out therein.

Thus, a tenant with repairing obligations under a lease or contract of business premises, may be liable under s4 if the repairs etc. are not done and especially if people are injured.

(4) Any reference in this section to a person having control of any premises or matter is a reference to a person having control of the premises or matter in connection with the carrying on by him of a trade, business or other undertaking (whether for profit or not).

S4(4)

This sub section makes it clear that 'landlords' etc. are those conducting a trade, business or other undertaking (whether for profit or not).

S4 does NOT cover those who are not in business as 'landlords' etc. This would probably exclude a person who had allowed someone else to use business premises for a purely philanthropic motive and where no consideration at all has passed (e.g. premises given 'free' to charitable undertakings).

Section 5

This section has been repealed but substantially re-enacted in the Environment Protection Act, 1990 which lays down duties on persons in control of premises in relation to harmful emissions into the atmosphere.

That Act should be referred to along with other environmental health measures as there is considerable overlap with this Act.

6 General duties of manufacturers etc as regards articles and substances for use at work

[(1) It shall be the duty of any person who designs, manufactures, imports or supplies any article for use at work or any article of fairground equipment–

(a) to ensure, so far as is reasonably practicable, that the article is so designed and constructed that it will be safe and without risks to health at all times when it is being set, used, cleaned or maintained by a person at work;

(b) to carry out or arrange for the carrying out of such testing and examination as may be necessary for the performance of the duty imposed on him by the preceding paragraph;

(c) to take such steps as are necessary to secure that persons supplied by that person with the article are provided with adequate information about the use for which the article is designed or has been tested and about any conditions necessary to ensure that it will be safe and without risks to health at all such times as are mentioned in paragraph (a) above and when it is being dismantled or disposed of; and

(d) to take such steps as are necessary to secure, so far as is reasonably practicable, that persons so supplied are provided with all such revisions of information provided to them by virtue of the preceding paragraph as are necessary by reason of its becoming known that anything gives rise to a serious risk to health or safety.

S6(1)

Duties are placed on designers, manufacturers, importers and suppliers in respect of articles for use at work to:

(a) ensure, so far as is reasonably practicable, that the article is so designed and constructed as to be safe and without risks to the health of persons at work when the article is being set, used, cleaned or maintained. This covers most operations NB. If foreign manufacturers etc. cannot be pursued in the British Courts, the section allows for actions to be taken against importers and suppliers.

(b) carry out or arrange for a competent person to carry out such testing and examination as may be necessary to comply with the duty in s6(1)(a). Most manufacturers etc. have research and development depts and quality control systems in place. If they do not or such systems fail there may be liability under this section although the words 'so far as is reasonably practicable' may provide a defence.

Manufactures etc. must certainly comply with the many DTI safety regulations and standards (i.e. CE marks) emanating from the EU which apply to particular articles nowadays.

(c) supply appropriate information about the article concerning:

its proper use

its testing

any conditions necessary to ensure that when in use it will be safe and without risks to health.

This information should certainly be supplied on packaging and enclosed with the article at work but also by means of manuals, technical data etc. Press advertisements may also be necessary to recall any faulty items.

These information provisions are reinforced by specific regulations which include:

PUWER 1998

Lifts Regulations, 1997

LOLER 1998

Definitions

'article for use at work' – see s53(1)

'risk' – see note to s3(2) above

'so far as is reasonably practicable' – see note to s2 above

'supply' – see s53(1)

'work' and 'at work' – see s52(1)

(1A) It shall be the duty of any person who designs, manufactures, imports or supplies any article of fairground equipment–

  (a) to ensure, so far as is reasonably practicable, that the article is so designed and constructed that it will be safe and without risks to health at all times when it is being used for or in connection with the entertainment of members of the public;

(b) to carry out or arrange for the carrying out of such testing and examination as may be necessary for the performance of the duty imposed on him by the preceding paragraph;

(c) to take such steps as are necessary to secure that persons supplied by that person with the article are provided with adequate information about the use for which the article is designed or has been tested and about any conditions necessary to ensure that it will be safe and without risks to health at all times when it is being used for or in connection with the entertainment of members of the public; and

(d) to take such steps as are necessary to secure, so far as is reasonably practicable, that persons so supplied are provided with all such revisions of information provided to them by virtue of the preceding paragraph as are necessary by reason of its becoming known that anything gives rise to a serious risk to health or safety.]

s6(1A)

Duties are placed on designers, manufacturers, importers and suppliers in respect of any article of fairground equipment to:

S6(1A)(a) ensure, so far as is reasonably practicable, that the article is so designed and constructed as to be safe and without risks to health at all times when it is used by members of the public for entertainment, etc. Designers etc. DO NOT seem to automatically have a defence if they can prove that somebody used the fairground article contrary to its normal use and there is an element of strict liability here.

NB. If foreign manufacturers etc. cannot be pursued in the British Courts, the section allows for actions to be taken against importers and suppliers.

S6(1A)(b)

This is in the same terms as s6(1)(b) except that it relates to entertainment etc. of members of the public – see notes above

S6(1A)(c)

This is in the same terms as s6(1)(c) except that it relates to entertainment etc. of members of the public – see notes above

S6(1A)(d)

This is in the same terms as s6(1)(d) except that it relates to entertainment etc. of members of the public – see notes above

Definitions

'fairground equipment' – see s53(1)

'risk' – see note to s3(2) above

'so far as is reasonably practicable' – see note to s2 above

'supply' – see s53(1)

(2) It shall be the duty of any person who undertakes the design or manufacture of any article for use at work [or of any article of fairground equipment] to carry out or arrange for the carrying out of any necessary research with a view to the discovery and, so far as is reasonably practicable, the elimination or minimisation of any risks to health or safety to which the design or article may give rise.

s6(2)

Designers and manufacturers (but NOT importers or suppliers) of articles for use at work (and fairground equipment) must carry out (or arrange for the carrying out) of proper research in order to discover and so far as is reasonably practicable eliminate or minimise any risks to health and safety to which the design or article or the way in which it is erected or installed may give rise.

Definitions
'article for use at work' – see s53(1)
'fairground equipment' – see s53(1)
'risk' – see note to s3(2) above
'so far as is reasonably practicable' – see note to s2 above

(3) It shall be the duty of any person who erects or installs any article for use at work in any premises where that article is to be used by persons at work [or who erects or installs any article of fairground equipment] to ensure, so far as is reasonably practicable, that nothing about the way in which [the article is erected or installed makes it unsafe or a risk to health at any such time as is mentioned in paragraph (a) of subsection (1) or, as the case may be, in paragraph (a) of subsection (1) or (1A) above].

s6(3)

Any person who erects or installs any article for use at work in any premises where it is to be used by persons at work or who erects or installs any article of fairground equipment must ensure, so far as is reasonably practicable, that the erection or installation is done properly so as to be safe and so as to minimise risks to health when properly used.
This subsection could apply to anyone who erects or installs such equipment, including employees, contractors, manufacturers, suppliers staff etc.
The erection, installation etc. must be at the times specified in s6(1)(a) and s6(A)(a)
Separate regulations apply to scaffolding for example.

Definitions
'article for use at work' – see s53(1)
'Fairground equipment – see s53(1)
'risk' – see note to s3(2) above
'so far as is reasonably practicable' – see note to s2 above
'work' and 'at work' – see s52(1)

[(4) It shall be the duty of any person who manufactures, imports or supplies any substance–

(a) to ensure, so far as is reasonably practicable, that the substance will be safe and without risks to health at all times when it is being used, handled, processed, stored or transported by a person at work or in premises to which section 4 above applies;

(b) to carry out or arrange for the carrying out of such testing and examination as may be necessary for the performance of the duty imposed on him by the preceding paragraph;

(c) to take such steps as are necessary to secure that persons supplied by that person with the substance are provided with adequate information about any risks to health or safety to which the inherent properties of the substance may give rise, about the results of any relevant tests which have been carried out on or in connection with the substance and about any conditions necessary to ensure that the substance will be safe and without risks to health at all such times as are mentioned in paragraph (a) above and when the substance is being disposed of; and

(d) to take such steps as are necessary to secure, so far as is reasonably practicable, that persons so supplied are provided with all such revisions of information provided to them by virtue of the preceding paragraph as are necessary by reason of its becoming known that anything gives rise to a serious risk to health or safety.]

6(4) DUTIES OF MANUFACTURERS, IMPORTERS AND SUPPLIERS OF ANY SUBSTANCE FOR USE AT WORK.

Duties are placed on, manufacturers, importers and suppliers (but not 'designers'/chemists') in respect of articles for use at work to:

(a) ensure, so far as is reasonably practicable, that the substance is safe and without risks to the health of persons at work when the substance is used, handled, processed, stored or transported or in premises covered by s4.

NB. If foreign manufacturers etc. cannot be pursued in the British Courts, the section allows for actions to be taken against importers and suppliers.

(b) carry out or arrange for a competent person to carry out such testing and examination as may be necessary to comply with the duty in s6(4)(a). Most manufacturers etc. have research and development depts and quality control systems in place. If they do not or such systems fail there may be liability under this section although the words 'so far as is reasonably practicable' may provide a defence.

(c) supply appropriate information about the substance concerning:
its proper use
its testing

any conditions necessary to ensure that when in use it will be safe and without risks to health.

This information should certainly be supplied on packaging and enclosed with the substance at work but also by means of data sheets, manuals, technical data etc. Press advertisements may also be necessary to warn of any problems.

These information provisions are reinforced by specific regulations which include:
COSHH 1999
CHIP 1999/2000
Control of Asbestos at work Regulations, 1987 (amended 1992)
Control of Lead at Work Regulations, 1998.

(5) It shall be the duty of any person who undertakes the manufacture of any [substance] to carry out or arrange for the carrying out of any necessary research with a view to the discovery and, so far as is reasonably practicable, the elimination or minimisation of any risks to health or safety to which the substance may give rise [at all such times as are mentioned in paragraph (a) of subsection (4) above].

s6(5)
Manufacturers (but NOT importers or suppliers or chemists) of substances for use at work must carry out (or arrange for the carrying out) of proper research in order to discover and so far as is reasonably practicable eliminate or minimise any risks to health and safety to which the substance may give rise.
This is reinforced by others laws (e.g. those on pharmacies and poisons)
Definitions
'substance' – see s53(1)
'substance for use at work' – see s53(1)
'risk' – see note to s3(2) above
'so far as is reasonably practicable' – see note to s2 above.

(6) Nothing in the preceding provisions of this section shall be taken to require a person to repeat any testing, examination or research which has been carried out otherwise than by him or at his instance, in so far as it is reasonable for him to rely on the results thereof for the purposes of those provisions.

S6(6)

This sub section makes it clear that in respect of any article or substance or any fairground equipment there is no obligation to re-test, re-examine or conduct further research where other persons have carried out such operations or the operations have been carried out by another at his behest PROVIDED it is reasonable for him to rely on those results.

It would NOT be reasonable to rely on such results, for example, if the other persons were subsequently found to be incompetent, unqualified etc. or information has subsequently come to light which brings into doubt the results. If accidents are occurring despite tests etc. this may be an indication that it is NOT reasonable to rely on results of tests etc. and re-testing is necessary.

(7) Any duty imposed on any person by any of the preceding provisions of this section shall extend only to things done in the course of a trade, business or other undertaking carried on by him (whether for profit or not) and to matters within his control.

S6(7)

The duties in ss6(1),(1A),(2),(3),(4),(5) and (6) are duties only in connection with things done in the course of a trade, business or other undertaking carried on by the person in question (whether profitable or not) and to matters within (not outside) his control.
Thus, it would appear that charities and other voluntary bodies are NOT covered by s6 PROVIDED they do not set out to conduct a business with a profit motive. They could be covered by other laws, however (e.g. negligence)
'Trade', 'business' and 'undertaking' are not defined by the Act but these terms have been widely defined in Common law, Company law, Partnership law, Tax law etc.
Basically, anyone conducting a business for profit whether they are a sole trader, partners, and Companies is covered whether they actually make a profit or not. Clubs etc. may also be covered if they have a profit motive.

(8) Where a person designs, manufactures, imports or supplies an article [for use at work or an article of fairground equipment and does so for or to another] on the basis of a written undertaking by that other to take specified steps sufficient to ensure, so far as is reasonably practicable, that the article will be safe and without risks to health [at all such times as are mentioned in paragraph (a) of subsection (1) or, as the case may be, in paragraph (a) of subsection (1) or (1A) above], the undertaking shall have the effect of relieving the first-mentioned person from the duty imposed [by virtue of that paragraph] to such extent as is reasonable having regard to the terms of the undertaking.

S6(8)

This sub section may provide a defence for certain designers, manufacturers, importers or suppliers of articles for use at work or articles of fairground equipment who do so for or to others on the basis of a written undertaking by those others to comply with s6(1)(a) or s6(1A)(a) insofar as it is reasonable for them to rely on the terms of the undertaking. Thus if, for example a manufacturer supplies a tower crane in kit form but takes a written undertaking from the Construction company they are supplying, that their expert will assemble the crane according to instructions provided and follow all appropriate safety procedures the manufacturer may escape s6 liability if the expert ignores the advice or is negligent. Indemnity terms in contracts are also advisable and may provide additional redress.

[(8A)   Nothing in subsection (7) or (8) above shall relieve any person who imports any article or substance from any duty in respect of anything which-

    (a) in the case of an article designed outside the United Kingdom, was done by and in the course of any trade, profession or other undertaking carried on by, or was within the control of, the person who designed the article; or

    (b) in the case of an article or substance manufactured outside the United Kingdom, was done by and in the course of any trade, profession or other undertaking carried on by, or was within the control of, the person who manufactured the article or substance.]

S6(8A)

The s6(7) and 6(8) defences do NOT extend to importers of articles or substances who are subject to the ss6(1) to 6(6) duties of anything which:
(a) regarding articles designed outside the UK, was done by and in the course of a business carried on by, or under the control of, the designer of the article; or
(b) regarding article or substance manufactured outside the UK, was done by and in the course of a business carried on by, or was within the control of, the manufacturer of the article or substance.

Thus, importers it seems may be prosecuted even if;
they are not conducting a business of importing (presumably this covers individuals)
matters are not under their control
undertakings have been given by others
other people are also at fault

(9)   Where a person ("the ostensible supplier") supplies any [article or substance] to another ("the customer") under a hire-purchase agreement, conditional sale agreement or credit-sale agreement, and the ostensible supplier-

(a) carries on the business of financing the acquisition of goods by others by means of such agreements; and

(b) in the course of that business acquired his interest in the article or substance supplied to the customer as a means of financing its acquisition by the customer from a third person ("the effective supplier"),

the effective supplier and not the ostensible supplier shall be treated for the purposes of this section as supplying the article or substance to the customer, and any duty imposed by the preceding provisions of this section on suppliers shall accordingly fall on the effective supplier and not on the ostensible supplier.

---

s6(9)

This sub section makes it clear that the original supplier (the 'effective supplier') of articles or substances to a customer may remain liable under s6 even if technically and legally the customer has a contract with the finance provider/creditor (the 'ostensible supplier').

In Hire Purchase, for example, if a A, Ltd inspects a van at the premises of B, Ltd (a van retailer) & A, Ltd agrees to take it on HP by finance provided by C, Ltd the following events occur:

B, Ltd will sell the van to C, Ltd who become the owner. This is a Sale of Goods Contract governed by the Sale of Goods Act, 1979 (see separate Point of Law book on this Act)

C, Ltd (the owner) will set up an HP contract with A, Ltd (the hirer)

C, Ltd remains the owner until all HP payments are made by A, Ltd and they exercise an option to become the purchasers. The hirer has NOT agreed to buy until this stage.

Technically there is no contract between A, Ltd and B, Ltd even though B, Ltd may supply the car to A Ltd.

The sub section makes it clear that B, Ltd would nevertheless have liability under s6 (cf. the position under the Consumer Credit Act 1974 where A Ltd's main remedies are against C, Ltd, although this Act allows for 'equal liability' in some circumstances).

'Credit sale' is similar to HP but the customer is the immediate owner of the goods. Even if the credit is given by another organisation the original supplier remains liable under s6 (cf. the situation under the Consumer Credit Act, 1974).

'Conditional sale' is another type of credit sale agreement but this time the supplier or creditor remains the owner until all instalments have been paid. The difference with HP is that in conditional sale the buyer has agreed to buy. Again the original supplier remains liable under s6.

NB. For more (including definitions) on this area of law consult the Consumer Credit Act, 1974 and books on Agency and Contract.

Definitions

For 'hire purchase', 'conditional sale' and 'credit sale' – see s53(1)

[(10)   For the purposes of this section an absence of safety or a risk to health shall be disregarded in so far as the case in or in relation to which it would arise is shown to be one the occurrence of which could not reasonably be foreseen; and in determining whether any duty imposed by virtue of paragraph (a) of subsection (1), (1A) or (4) above has been performed regard shall be had to any relevant information or advice which has been provided to any person by the person by whom the article has been designed, manufactured, imported or supplied or, as the case may be, by the person by whom the substance has been manufactured, imported or supplied.]

S6(10)

This sub-section provides for certain defences to an action under s6(1)(1A) or 6(4) as follows:
Occurrences which could not reasonably be foreseen (i.e. remoteness of damage, Act of God, inevitable accident, etc) – e.g. unusual weather conditions, disasters etc.,
regard being paid to any relevant information or advice provided to any person by designers, manufacturers, importers or suppliers of articles or manufacturers, importers or suppliers of substances. Thus, if people ignore data sheets etc. supplied by manufacturers etc., the manufacturer etc may have a defence.

7   General duties of employees at work

It shall be the duty of every employee while at work-

   (a) to take reasonable care for the health and safety of himself and of other persons who may be affected by his acts or omissions at work; and

Section 7 – General Duties of employees at work.
(a) This subsection obliges employees NOT to be negligent in the performance of any tasks at work. If the employee is negligent and injures himself or others he may be prosecuted and fined up to 5000 pounds in a magistrates court and an unlimited fine in a Crown Court. This may be in addition to fines on other responsible persons under ss 2 to 6 above.

   (b) as regards any duty or requirement imposed on his employer or any other person by or under any of the relevant statutory provisions, to co-operate with him so far as is necessary to enable that duty or requirement to be performed or complied with.

(b) This subsection obliges the employee to co-operate with the employer so as not to obstruct him in the carrying out of his duties under health and safety legislation. Again for breach the employee can be fined up to 5000 pounds in a magistrates court and an unlimited fine in the Crown Court in addition to other persons' liability under ss2 to 6 above.

These employees duties are expanded upon in the Management of Health & safety at Work Regulations,1999 and almost every other set of regulations will lay down employees duties too.

Definitions:
'Employee' – see s53(1)
'Employer' – see s53(1)
'the relevant statutory provisions' – see s53(1)
'at work' – see s52(1).

The term 'employee' can include ordinary operatives, supervisors, line managers, senior management and even managing directors and chief executives of Companies. All can be held liable under s7.

8   Duty not to interfere with or misuse things provided pursuant to certain provisions

No person shall intentionally or recklessly interfere with or misuse anything provided in the interests of health, safety or welfare in pursuance of any of the relevant statutory provisions.

Section 8 – Duty not to interfere with or misuse things provided pursuant to certain statutory provisions.

This section is often,erroneously,thought to apply to employees only. It actually applies to any person and places on them a duty not to intentionally or recklessly interfere with or misuse anything provided as a warning or preventive measure by virtue of health & safety or welfare legislation. A 5000 pounds fine can be levied in the magistrates court for breach or an unlimited fine on the Crown Court. Criminal damage actions could be brought under other legislation.

The section applies to defacement or destruction of warning notices under the Health & Safety (Signs and Signals) Regulations 1996 and to fire signs provided pursuant to fire risk assessments under the Workplace (Fire Precautions) Regulations 1997 but also to PPE and any safety device provided there is a statutory duty to provide it.

'Relevant statutory provisions' – see s53(1)

9 Duty not to charge employees for things done or provided pursuant to certain specific requirements

No employer shall levy or permit to be levied on any employee of his any charge in respect of anything done or provided in pursuance of any specific requirement of the relevant statutory provisions.

Section 9 – Duty not to charge employees for things done or provided pursuant to certain specific requirements.

This section makes it clear that an employer must NOT charge his employees for any personal protective equipment or other things provided by virtue of a specific statutory duty. Thus, if it is LAW that some protective equipment, device etc. is to be provided the employer must provide it at his cost and NOT seek to recover the cost from employees. This is backed up by the Personal Protective Equipment Regulations, 1992 whereby PPE has to be provided if an employer is not protecting employees by better control/preventive measures Essential Glasses for users of display screen equipment must also be provided free to employees. These other regulations lay down requirements as to compliance with EU standards, maintenance, storage, replacement etc.

'Employee' – see s53(1)
'Employer' – see s53(1) (NB: These can be sole traders, partners, unincorporated associations or Companies)
'the relevant statutory provisions' – see s53(1)

Breach of this section is punishable by up to a 5000 pound fine in a magistrates court or an unlimited fine in the Crown Court.

THE HEALTH AND SAFETY COMMISSION AND THE HEALTH AND SAFETY EXECUTIVE

10 Establishment of the Commission and the Executive

(1) There shall be two bodies corporate to be called the Health and Safety Commission and the Health and Safety Executive which shall be constituted in accordance with the following provisions of this section.

The Health & Safety Commission and the Health & safety Executive.

(1) This subsection establishes as corporate bodies the Health & Safety Commission (HSC) and Health & Safety Executive (HSE). The former is a policy making body and the latter is an enforcement body. Both are QUANGOs (Quasi Autonomous Non governmental organisations), although the Dept for Education & Employment keeps close control. Most previous inspectorates are now merged into the HSE.

(2) The Health and Safety Commission (hereafter in this Act referred to as "the Commission") shall consist of a chairman appointed by the Secretary of State and not less than six nor more than nine other members appointed by the Secretary of State in accordance with subsection (3) below.

(2) The HSC is made up of:

A Chairperson – appointed by Secretary of State for Education & Employment (SoS) Between 6 and 9 other members appointed by SoS.

(3) Before appointing the members of the Commission (other than the chairman) the Secretary of State shall–

(a) as to three of them, consult such organisations representing employers as he considers appropriate;

(b) as to three others, consult such organisations representing employees as he considers appropriate; and

(c) as to any other members he may appoint, consult such organisations representing local authorities and such other organisations, including professional bodies, the activities of whose members are concerned with matters relating to any of the general purposes of this Part, as he considers appropriate.

(3) Before appointing the 6-9 members referred to above the SoS must consult Employers organisations (e.g. CBI) as to 3 members; employees organisations (e.g. TUC) regarding 3 members and local authority associations (e.g. ADCs) and health & safety professional bodies (e.g. CIEH).

(4) The Secretary of State may appoint one of the members to be deputy chairman of the Commission.

(4) Anyone of the above can be appointed deputy chairperson by the SoS.

(5) The Health and Safety Executive (hereafter in this Act referred to as "the Executive") shall consist of three persons of whom one shall be appointed by the Commission with the approval of the Secretary of State to be the director of the Executive and the others shall be appointed by the Commission with the like approval after consultation with the said director.

(5) The HSE consists of three persons as follows:

Director General of HSE (DG) – appointed by HSC with SoS approval

Two other persons – appointed by HSC with SoS approval after consultation with the DG.

(6) The provisions of Schedule 2 shall have effect with respect to the Commission and the Executive.

(6) Schedule 2 of the Act lays down detailed rules (rather like the rules of a company) concerning the working of the HSC and HSE. (see Schedule 2 post)

(7) The functions of the Commission and of the Executive, and of their officers and servants, shall be performed on behalf of the Crown.

(7) The functions of HSC and HSE are exercised on behalf of the Crown. They are thus protected from criminal prosecution. Both bodies, although QUANGOs, are staffed by civil servants.

[(8) For the purposes of any civil proceedings arising out of those functions, the Crown Proceedings Act 1947 and the Crown Suits (Scotland) Act 1857 shall apply to the Commission and the Executive as if they were government departments within the meaning of the said Act of 1947 or, as the case may be, public departments within the meaning of the said Act of 1857.]

(8) Although HSC and HSE are Crown bodies they can be proceeded against in civil law in the same way as any other government department since 1947 (or in Scotland since 1857).

Definitions:

'Employer' – see s53(1)

'Employee' – see s53(1)

'local authorities' – see s53(1)

11 General functions of the Commission and the Executive

(1) In addition to the other functions conferred on the Commission by virtue of this Act, but subject to subsection (3) below, it shall be the general duty of the Commission to do such things and make such arrangements as it considers appropriate for the general purposes of this Part ...

> **Section 11 – General Functions of the Commission and Executive.**
> *(1) The HSC as a 'creature of Statute' only has powers under this Act. It also has all incidental powers and powers to make appropriate arrangements to implement Part 1 of the Act by virtue of this sub section.*

(2) It shall be the duty of the Commission ...–

   (a) to assist and encourage persons concerned with matters relevant to any of the general purposes of this Part to further those purposes;

   (b) to make such arrangements as it considers appropriate for the carrying out of research, the publication of the results of research and the provision of training and information in connection with those purposes, and to encourage research and the provision of training and information in that connection by others;

   (c) to make such arrangements as it considers appropriate for securing that government departments, employers, employees, organisations representing employers and employees respectively, and other persons concerned with matters relevant to any of those purposes are provided with an information and advisory service and are kept informed of, and adequately advised on, such matters;

   (d) to submit from time to time to the authority having power to make regulations under any of the relevant statutory provisions such proposals as the Commission considers appropriate for the making of regulations under that power.

> *(2) The duties of HSC are to:*
> *(a) assist and encourage others to further the purposes of the Act;*
> *(b) carry out research and publish results and training an information in connection with that research;*
> *(c) provide an information and advisory service to a wide cross section of organisations and persons.*
> *(d) submit to the SoS proposals for making new Regulations under the Act.*

(3) It shall be the duty of the Commission–

    (a) to submit to the Secretary of State from time to time particulars of what it proposes to do for the purpose of performing its functions; and

    (b) subject to the following paragraph, to ensure that its activities are in accordance with proposals approved by the Secretary of State; and

    (c) to give effect to any directions given to it by the Secretary of State.

> *(3) Further duties on the HSC are to:*
> *(a) submit proposals to the SoS for the performance of its functions*
> *(b) ensure its activities are in accordance with proposals approved by the SoS*
> *(c) give effect to any directions given by the SoS.*
> *There is, thus, a lot of central control.*

(4) In addition to any other functions conferred on the Executive by virtue of this Part, it shall be the duty of the Executive–

    (a) to exercise on behalf of the Commission such of the Commission's functions as the Commission directs it to exercise; and

    (b) to give effect to any directions given to it by the Commission otherwise than in pursuance of paragraph (a) above;

but, except for the purpose of giving effect to directions given to the Commission by the Secretary of State, the Commission shall not give to the Executive any directions as to the enforcement of any of the relevant statutory provisions in a particular case.

> *(4) In addition to other functions under the Act the HSE has the following duties:*
> *(a) to act as a delegate for the HSC as and when HSC directs*
> *(b) to give effect to any HSC directions other than (a).*
>
> *But as regards SoS directions to HSC, the HSC may NOT give the HSE any directions regarding the enforcement of statutory health & safety provisions in particular cases.*

(5) Without prejudice to subsection (2) above, it shall be the duty of the Executive, if so requested by a Minister of the Crown–

    (a) to provide him with information about the activities of the Executive in connection with any matter with which he is concerned; and

    (b) to provide him with advice on any matter with which he is concerned on which relevant expert advice is obtainable from any of the officers or servants of the Executive but which is not relevant to any of the general purposes of this Part.

(5) The HSE is duty bound if requested by a Minister of the Crown to:
(a) provide him with information on HSE activities he is concerned about; and
(b) provide him with advice on any matter he is concerned with on which relevant expert advice is obtainable from HSE officers or servant regardless of whether it is covered by Part 1 of the Act.
Thus, HSE are advisors and providers of information to any Government Minister who request it.

(6) The Commission and the Executive shall, subject to any directions given to it in pursuance of this Part, have power to do anything (except borrow money) which is calculated to facilitate, or is conducive or incidental to, the performance of any function of the Commission or, as the case may be, the Executive (including a function conferred on it by virtue of this subsection).

(6) Both HSC and HSE have all incidental powers (e.g. to hire and fire staff, maintain offices etc.) to enable them to exercise their statutory powers. This is, however, subject to any directions given to them under Part 1 (e.g. by SoS) and
a prohibition on borrowing money. The Government finances both bodies and keeps financial control which impairs their autonomy somewhat.

'Employer' – see s53(1)
'Employee' – see s53(1)
'Relevant statutory provisions' – see s53(1).

12 Control of the Commission by the Secretary of State

The Secretary of State may–

(a) approve, with or without modifications, any proposals submitted to him in pursuance of section 11(3)(a);

(b) give to the Commission at any time such directions as he thinks fit with respect to its functions (including directions modifying its functions, but not directions conferring on it functions other than any of which it was deprived by previous directions given by virtue of this paragraph), and any directions which it appears to him requisite or expedient to give in the interests of the safety of the State.

Section 12 Control of the Commission by the Secretary of State.
The SoS may:
(a) approve HSC proposals under s11(3)(a) with or without modifications
The SOS may be subject to EU law, however)

(b) give HSC at any time such directions as he thinks fit with respect to HSC functions (including directions modifying its functions but NOT giving it new functions unless they were previously taken away by direction of the SOS under s12. The SOS may also give directions to HSC in the interests of the safety of the State.

13 Other powers of the Commission

(1) The Commission shall have power–

(a) to make agreements with any government department or other person for that department or person to perform on behalf of the Commission or the Executive (with or without payment) any of the functions of the Commission or, as the case may be, of the Executive;

(b) subject to subsection (2) below, to make agreements with any Minister of the Crown, government department or other public authority for the Commission to perform on behalf of that Minister, department or authority (with or without payment) functions exercisable by the Minister, department or authority (including, in the case of a Minister, functions not conferred by an enactment), being functions which in the opinion of the Secretary of State can appropriately be performed by the Commission in connection with any of the Commission's functions;

(c) to provide (with or without payment) services or facilities required otherwise than for the general purposes of this Part in so far as they are required by any government department or other public authority in connection with the exercise by that department or authority of any of its functions;

(d) to appoint persons or committees of persons to provide the Commission with advice in connection with any of its functions and (without prejudice to the generality of the following paragraph) to pay to persons so appointed such remuneration as the Secretary of State may with the approval of the Minister for the Civil Service determine;

(e) in connection with any of the functions of the Commission, to pay to any person such travelling and subsistence allowances and such compensation for loss of remunerative time as the Secretary of State may with the approval of the Minister for the Civil Service determine;

(f) to carry out or arrange for or make payments in respect of research into any matter connected with any of the Commission's functions, and to disseminate or arrange for or make payments in respect of the dissemination of information derived from such research;

(g) to include, in any arrangements made by the Commission for the provision of facilities or services by it or on its behalf, provision for the making of payments to the Commission or any person acting on its behalf by other parties to the arrangements and by persons who use those facilities or services.

(2) Nothing in subsection (1)(b) shall authorise the Commission to perform any function of a Minister, department or authority which consists of a power to make regulations or other instruments of a legislative character.

> **Section 13 – Other powers of the Commission**
> *This section gives power to the HSC to make agreements and to appoint and pay persons to help carry out the functions of the HSC or HSE. The powers are fairly self explanatory.*

14 Power of the Commission to direct investigations and inquiries

(1) This section applies to the following matters, that is to say any accident, occurrence, situation or other matter whatsoever which the Commission thinks it necessary or expedient to investigate for any of the general purposes of this Part or with a view to the making of regulations for those purposes; and for the purposes of this subsection it is immaterial whether the Executive is or is not responsible for securing the enforcement of such (if any) of the relevant statutory provisions as relate to the matter in question.

> **Section 14 – Power of the Commission to direct investigations and inquiries**
> *(1) This subsection empowers the HSC to conduct investigations and inquiries into accidents, occurences and other situations for the purposes of Part 1 of the Act or with a view to the making of new Regulations even if HSE lacks the power to enforce the matters in question for the time being.*
> *This section has been used to set up inquiries etc. into the many major disasters which have happened since the passing of the Act (e.g. Kings Cross fire, Clapham Rail Disaster, Piper Alpha etc. etc.).*

(2) The Commission may at any time–

(a) direct the Executive or authorise any other person to investigate and make a special report on any matter to which this section applies; or

(b) with the consent of the Secretary of State direct an inquiry to be held into any such matter;

(2) HSC may direct the HSE or authorise any other person to investigate and make a special report on any matter to which this section applies and may with SOS consent direct an inquiry to be held into any such matter.

(3) Any inquiry held by virtue of subsection (2)(b) above shall be held in accordance with regulations made for the purposes of this subsection by the Secretary of State, and shall be held in public except where or to the extent that the regulations provide otherwise.

(3) Any inquiry under s14(2)(b) shall be held in accordance with regulations made by the SoS, and held in public unless the Regulations direct otherwise.
The Regulations have been made, i.e. The Health & Safety Inquiries (Procedure) Regulations, 1975 (SI 1975/335)

(4) Regulations made for the purposes of subsection (3) above may in particular include provision-

   (a) conferring on the person holding any such inquiry, and any person assisting him in the inquiry, powers of entry and inspection;

   (b) conferring on any such person powers of summoning witnesses to give evidence or produce documents and power to take evidence on oath and administer oaths or require the making of declarations;

   (c) requiring any such inquiry to be held otherwise than in public where or to the extent that a Minister of the Crown so directs.

(4) Regulations under s14(3) may include provision for:
(a) conferring powers of entry and inspection
(b) summoning witnesses, taking statements on oath & the administration of oaths or statutory declarations
(c) inquiries to be held in private if a Crown Minister so directs.

(5) In the case of a special report made by virtue of subsection (2)(a) above or a report made by the person holding an inquiry held by virtue of subsection (2)(b) above, the Commission may cause the report, or so much of it as the Commission thinks fit, to be made public at such time and in such manner as the Commission thinks fit.

(5) Special Reports under s14(2)(a) or Reports of inquiries under s14(2)(b) may be published by HSC in whole or part and at such times and in such manner as HSC thinks fit.

(6) The Commission–

    (a) in the case of an investigation and special report made by virtue of subsection (2)(a) above (otherwise than by an officer or servant of the Executive), may pay to the person making it such remuneration and expenses as the Secretary of State may, with the approval of the Minister for the Civil Service, determine;

    (b) in the case of an inquiry held by virtue of subsection (2)(b) above, may pay to the person holding it and to any assessor appointed to assist him such remuneration and expenses, and to persons attending the inquiry as witnesses such expenses, as the Secretary of State may, with the like approval, determine; and

    (c) may, to such extent as the Secretary of State may determine, defray the other costs, if any, of any such investigation and special report or inquiry.

(6) This entitles HSC to pay persons conducting inquiries and making reports and to defray other costs.

(7) Where an inquiry is directed to be held by virtue of subsection (2)(b) above into any matter to which this section applies arising in Scotland, being a matter which causes the death of any person, no inquiry with regard to that death shall, unless the Lord Advocate otherwise directs, be held in pursuance of the [Fatal Accidents and Sudden Deaths Inquiry (Scotland) Act 1976].

(7) This subsection applies to Scotland only and requires that inquiries under s14(2)(b) regarding deaths shall be held in pursuance of the Fatal Accidents Inquiry (Scotland) Act, 1895 unless the Lord Advocate otherwise directs.

HEALTH AND SAFETY REGULATIONS AND APPROVED CODES OF PRACTICE

15 Health and safety regulations

[(1) Subject to the provisions of section 50, the Secretary of State, the Minister of Agriculture, Fisheries and Food or the Secretary of State and that Minister acting jointly shall have power to make regulations under this section for any of the general purposes of this Part (and regulations so made are in this Part referred to as "health and safety regulations").]

Health & safety Regulations and Approved Codes of Practice

(1) This subsection empowers (subject to s50), the SoS, the Minister of Agriculture, Fisheries and Food (MAFF)* or both of them jointly to make Health & Safety Regulations for the purposes of Part 1.

This is a very important power and many Regulations have now been passed under the Act (see non exhaustive list in the Introduction). All regulations are now made under this provision instead of older legislation and, with a few exceptions, made applicable to all workplaces within Great Britain. EU Directives will, of course, oblige the SOS to exercise his powers to comply with EU law and most recent Regulations have come from the EU.

NB* MAFF now has a different title but the powers are the same.

(2) Without prejudice to the generality of the preceding subsection, health and safety regulations may for any of the general purposes of this Part make provision for any of the purposes mentioned in Schedule 3.

(2) Health & safety regulations may make provision for any of the purposes in Schedule 3 (see post). The powers are wide and fairly self explanatory.

(3) Health and safety regulations–

    (a) may repeal or modify any of the existing statutory provisions;

    (b) may exclude or modify in relation to any specified class of case any of the provisions of sections 2 to 9 or any of the existing statutory provisions;

    (c) may make a specified authority or class of authorities responsible, to such extent as may be specified, for the enforcement of any of the relevant statutory provisions.

(3) Health & safety Regulations may:
(a) repeal or modify any existing statutory provisions
(b) may exclude or modify s2 to 9 or any existing statutory provisions in respect of specified classes of case.
(c) may make specified authorities (or classes) responsible to such extent as specified in the regulations for enforcement of any of the relevant statutory provisions.

Definitions;
'existing statutory provisions' – see s53(1) and Schedule 1 column 3 (i.e. older legislation and regulations made under their provisions)
'relevant statutory provisions' – see s53(1)
'modify' – see s82(1)(c)

(4) Health and safety regulations–

    (a) may impose requirements by reference to the approval of the Commission or any other specified body or person;

    (b) may provide for references in the regulations to any specified document to operate as references to that document as revised or re-issued from time to time.

(4) Health & safety regulations
(a) may impose requirements requiring approval by the HSC or any other specified body or person
(b) may provide for references in the Regulations to specified documents as revised or reissued from time to time.
An example of this is Guidance Note EH40 on Occupational Exposure Limits for Chemicals issued annually but referred to in the COSHH Regulations 1999. New Regulations do NOT have to be made every time EH40 is revised; the power to refer to the updated document is contained in the 1999 regulations.

(5) Health and safety regulations–

    (a) may provide (either unconditionally or subject to conditions, and with or without limit of time) for exemptions from any requirement or prohibition imposed by or under any of the relevant statutory provisions;

    (b) may enable exemptions from any requirement or prohibition imposed by or under any of the relevant statutory provisions to be granted (either unconditionally or subject to conditions, and with or without limit of time) by any specified person or by any person authorised in that behalf by a specified authority.

(5) Health & safety Regulations may
(a) provide, conditionally or unconditionally & without time limit, for exemptions from any requirement or prohibition of the relevant statutory provisions (see s53(1))
(b) provide for any person authorised by a specified authority to exempt from any requirement or prohibition under any of the relevant statutory provisions, either conditionally or unconditionally & without time limit.

(6) Health and safety regulations–

    (a) may specify the persons or classes of persons who, in the event of a contravention of a requirement or prohibition imposed by or under the regulations, are to be guilty of an offence, whether in addition to or to the exclusion of other persons or classes of persons;

(b) may provide for any specified defence to be available in proceedings for any offence under the relevant statutory provisions either generally or in specified circumstances;

(c) may exclude proceedings on indictment in relation to offences consisting of a contravention of a requirement or prohibition imposed by or under any of the existing statutory provisions, sections 2 to 9 or health and safety regulations;

(d) may restrict the punishments [(other than the maximum fine on conviction on indictment)] which can be imposed in respect of any such offence as is mentioned in paragraph (c) above;

[(e) in the case of regulations made for any purpose mentioned in section 1(1) of the Offshore Safety Act 1992, may provide that any offence consisting of a contravention of the regulations, or of any requirement or prohibition imposed by or under them, shall be punishable on conviction on indictment by imprisonment for a term not exceeding two years, or a fine, or both.]

(6) Health & safety Regulations may–

(a) specify who is to be regarded as liable to criminal prosecution if the Regulations are breached & whether this is in addition to or exclusion of others.
This usually covers the 'employer' and self employed persons' but can include others (e.g. employee, competent persons, etc etc.).

(b) include provision for defences to criminal charges (e.g. 'due diligence', 'so far as is reasonably practicable' etc.)

(c) exclude Crown Court proceedings in ss 2 to 9, any existing statutory provisions or any relevant statutory provisions.

(d) restrict criminal punishments (other than maximum fines in the Crown Court) in respect of offences mentioned in s15(6)(c).

(e) In the case of Regulations made under the Offshore Safety Act, 1992 may provide for criminal penalties in the Crown Court of 2 years imprisonment, or a fine or both.

(7) Without prejudice to section 35, health and safety regulations may make provision for enabling offences under any of the relevant statutory provisions to be treated as having been committed at any specified place for the purpose of bringing any such offence within the field of responsibility of any enforcing authority or conferring jurisdiction on any court to entertain proceedings for any such offence.

(7) Subject to s35 (rules concerning court venue), health & safety regulations may make provision for allowing offences to be treated as having been committed at any specified place for the purpose of bringing court proceedings or for increasing the powers of enforcement authorities.
Thus, usual rules as to venue may be waived by regulations in appropriate cases.

(8) Health and safety regulations may take the form of regulations applying to particular circumstances only or to a particular case only (for example, regulations applying to particular premises only).

> *(8) Health and Safety Regulations may be enacted to apply to particular circumstances or cases only, rather than generally.*

(9) If an Order in Council is made under section 84 (3) providing that this section shall apply to or in relation to persons, premises or work outside Great Britain then, notwithstanding the Order, health and safety regulations shall not apply to or in relation to aircraft in flight, vessels, hovercraft or offshore installations outside Great Britain or persons at work outside Great Britain in connection with submarine cables or submarine pipelines except in so far as the regulations expressly so provide.

> *(9) Orders in (Privy) Council may be made under s84(3) in relation to work outside GB on oil and gas rigs mainly. This section makes it clear that such Orders do NOT apply to aircraft in flight, vessels, hovercraft, or offshore installations outside GB or in connection with submarine cables or pipelines except so far as the Regulations EXPRESSLY provide. Thus, it will NOT be readily IMPLIED that such Regulations apply outside GB unless expressly stated.*

(10) In this section "specified" means specified in health and safety regulations.

> *(10) 'specified' wherever it is mentioned in s15 means specified in health & safety regulations. No other source will do.*
>
> *NB. There are additional powers to make regulations under s15 under the Offshore Safety Act, 1992 and the Railways Act,1993.*
>
> *Ministers must keep their law making powers within those granted by Parliament under s15. If they do not they will be acting ultra vires, illegally and are open to challenge in the courts by judicial review.*

16 Approval of codes of practice by the Commission

(1) For the purpose of providing practical guidance with respect to the requirements of any provision of sections 2 to 7 or of health and safety regulations or of any of the existing statutory provisions, the Commission may, subject to the following subsection–

    (a) approve and issue such codes of practice (whether prepared by it or not) as in its opinion are suitable for that purpose;

    (b) approve such codes of practice issued or proposed to be issued otherwise than by the Commission as in its opinion are suitable for that purpose.

Section 16 – Approval of Codes of Practice by the Commission.

(1) The HSC may for the purpose of providing practical guidance on the requirements of ss2 to 9, or any health & safety regulations or any existing statutory provisions (see s53(1)) approve and issue Approved Codes of Practice (ACOPs) which in its opinion are suitable for the purpose whether prepared by the HSC or not.

NB. Many such ACOPS have now been approved & are published with the appropriate Regulations. They are listed in HSE Books booklists and available from them.

(2) The Commission shall not approve a code of practice under subsection (1) above without the consent of the Secretary of State, and shall, before seeking his consent, consult–

    (a) any government department or other body that appears to the Commission to be appropriate (and, in particular, in the case of a code relating to electro-magnetic radiations, the National Radiological Protection Board); and

    (b) such government departments and other bodies, if any, as in relation to any matter dealt with in the code, the Commission is required to consult under this section by virtue of directions given to it by the Secretary of State.

(2) HSC approval of ACOPS is subject to SOS consent & before seeking that consent HSC must consult any government department or other appropriate body (including the National Radiological Protection Board in relation to ACOPs relating to electro-magnetic radiation) and other government departments and bodies the HSC is directed to consult by the SOS.

(3) Where a code of practice is approved by the Commission under subsection (1) above, the Commission shall issue a notice in writing–

    (a) identifying the code in question and stating the date on which its approval by the Commission is to take effect; and

    (b) specifying for which of the provisions mentioned in subsection (1) above the code is approved.

(3) Where the HSC approves an ACOP, HSC must issue a written notice:

(a) identifying the ACOP in question & stating the commencement date; and

(b) specifying which Act or Regulation the ACOP relates to (e.g. ACOP regarding Chemicals relates to the COSHH Regulations, 1999).

(4) The Commission may–

    (a) from time to time revise the whole or any part of any code of practice prepared by it in pursuance of this section;

(b) approve any revision or proposed revision of the whole or any part of any code of practice for the time being approved under this section;

and the provisions of subsections (2) and (3) above shall, with the necessary modifications, apply in relation to the approval of any revision under this subsection as they apply in relation to the approval of a code of practice under subsection (1) above.

(4) HSC may periodically revise the whole or any part of an ACOP to approve any revision or proposed revision of the whole or part of an ACOP but these are subject to the consultation and consent requirements of s16(2) and the notice requirements of s16(3).

(5) The Commission may at any time with the consent of the Secretary of State withdraw its approval from any code of practice approved under this section, but before seeking his consent shall consult the same government departments and other bodies as it would be required to consult under subsection (2) above if it were proposing to approve the code.

(5) HSC may withdraw an ACOP with SOS approval but prior to this must consult under s16(2) as if it were approving an ACOP.

(6) Where under the preceding subsection the Commission withdraws its approval from a code of practice approved under this section, the Commission shall issue a notice in writing identifying the code in question and stating the date on which its approval of it is to cease to have effect.

(6) If an ACOP is withdrawn under s 16(5), the HSC shall issue a written notice identifying the ACOP and stating the date of cessation.

(7) References in this part to an approved code of practice are references to that code as it has effect for the time being by virtue of any revision of the whole or any part of it approved under this section.

(7) References to an ACOP are deemed to be references to the current ACOP, whether revised in whole or in part.

(8) The power of the Commission under subsection (1)(b) above to approve a code of practice issued or proposed to be issued otherwise than by the Commission shall include power to approve a part of such a code of practice; and accordingly in this Part "code of practice" may be read as including a part of such a code of practice.

(8) HSC's power under s16(1)(b) to approve an ACOP issued otherwise than by the HSC shall include power to approved the CoP in part only.

Definitions;

'code of practice' – see s53(1)

NB. Guidance notes are also issued by HSE but have no statutory equivalent to s16 or 17. They are merely suggestions as to how the Regulations may be implemented but there may be other methods of compliance. They may have some relevance to the civil law also.

17   Use of approved codes of practice in criminal proceedings

Section 17 – Use of Approved Codes of Practice in criminal proceedings.

This section makes clear the status of ACOPs.

(1)   A failure on the part of any person to observe any provision of an approved code of practice shall not of itself render him liable to any civil or criminal proceedings; but where in any criminal proceedings a party is alleged to have committed an offence by reason of a contravention of any requirement or prohibition imposed by or under any such provision as is mentioned in section 16(1) being a provision for which there was an approved code of practice at the time of the alleged contravention, the following subsection shall have effect with respect to that code in relation to those proceedings.

(1)   A failure to observe an ACOP does not of itself make anyone liable to civil or criminal court proceedings.

However, if in criminal proceedings a person is charged with breach of ss 2 to 7, or of health & safety regulations or existing statutory provisions & there is an ACOP relating to those provisions then s17(2) shall have effect.

(2)   Any provision of the code of practice which appears to the court to be relevant to the requirement or prohibition alleged to have been contravened shall be admissible in evidence in the proceedings; and if it is proved that there was at any material time a failure to observe any provision of the code which appears to the court to be relevant to any matter which it is necessary for the prosecution to prove in order to establish a contravention of that requirement or prohibition, that matter shall be taken as proved unless the court is satisfied that the requirement or prohibition was in respect of that matter complied with otherwise than by way of observance of that provision of the code.

(2) If in criminal proceedings to which s17(1) is relevant, there is an ACOP, that ACOP is admissible in evidence in the proceedings. If the breach of ACOP is proved at any material time, breach of the appropriate provision is taken to be proved, UNLESS THE COURT IS SATISFIED THAT THE DEFENDANT COMPLIED IN SOME OTHER WAY WHICH IS AT LEAST AS GOOD AS THE PROVISIONS OF THE CODE.

Thus, Employers etc. are advised to follow appropriate ACOPs but if they have set up better methods than the ACOP they will be protected from a successful prosecution. Conversely, if the ACOP has been ignored or his own measures are NOT as good as the ACOP, the employer will be liable for breach of the appropriate provision.

ACOPS, therefore, whilst NOT law, may be used as proof that an employer etc. did not do what a law required whether under the ACOP or his own better provisions.

(3) In any criminal proceedings–

(a) a document purporting to be a notice issued by the Commission under section 16 shall be taken to be such a notice unless the contrary is proved; and

(b) a code of practice which appears to the court to be the subject of such a notice shall be taken to be the subject of that notice unless the contrary is proved.

(3) In any criminal proceedings any document which appears to be an HSC notice under s16 shall be taken as a valid notice and any code of practice which appears to the court to be the subject of such notice shall be taken as valid unless the contrary is proved. This saves people attacking the procedural aspects and wasting the court's time.

Enforcement

18 Authorities responsible for enforcement of the relevant statutory provisions

(1) It shall be the duty of the Executive to make adequate arrangements for the enforcement of the relevant statutory provisions except to the extent that some other authority or class of authorities is by any of those provisions or by regulations under subsection (2) below made responsible for their enforcement.

Section 18 – Authorities Responsible for enforcement of the relevant statutory provisions.

(1) The HSE are the main enforcement authority for the Act and all Regulations made under the Act but other authorities may act under relevant Regulations, eg. Environmental Health Departments of Local District Councils enforce in the service sector and in shops and offices etc.

(2) The Secretary of State may by regulations–

   (a) make local authorities responsible for the enforcement of the relevant statutory provisions to such extent as may be prescribed;

   (b) make provision for enabling responsibility for enforcing any of the relevant statutory provisions to be, to such extent as may be determined under the regulations–

       (i) transferred from the Executive to local authorities or from local authorities to the Executive; or

       (ii) assigned to the Executive or to local authorities for the purpose of removing any uncertainty as to what are by virtue of this subsection their respective responsibilities for the enforcement of those provisions;

and any regulations made in pursuance of paragraph (b) above shall include provision for securing that any transfer or assignment effected under the regulations is brought to the notice of persons affected by it.

(2) This empowers the SOS to make regulations making local authorities liable for enforcement in some areas or transferring HSE powers to local authorities or vice versa PROVIDED that all affected by the change are given notice of it.

(3) Any provision made by regulations under the preceding subsection shall have effect subject to any provision made by health and safety regulations ... in pursuance of section 15(3)(c).

(3) The power to make regulations under s18(2) is inferior to the power to make health & safety regulations under s15(3)(c). The latter prevails in the event of dispute regarding interpretation.

(4) It shall be the duty of every local authority–

(a) to make adequate arrangements for the enforcement within their area of the relevant statutory provisions to the extent that they are by any of those provisions or by regulations under subsection (2) above made responsible for their enforcement; and

(b) to perform the duty imposed on them by the preceding paragraph and any other functions conferred on them by any of the relevant statutory provisions in accordance with such guidance as the Commission may give them.

(4) Local Authorities if given power under s18(2) must make adequate arrangements for the enforcement of the Act and all health & safety regulations entrusted to it. This is usually done by Environmental Health Inspectors. They must also perform their duties and functions in compliance with any guidance by the HSC.

(5) Where any authority other than ... the Executive or a local authority is by any of the relevant statutory provisions or by regulations under subsection (2) above made responsible for the enforcement of any of those provisions to any extent, it shall be the duty of that authority–

(a) to make adequate arrangements for the enforcement of those provisions to that extent; and

(b) to perform the duty imposed on the authority by the preceding paragraph and any other functions conferred on the authority by any of the relevant statutory provisions in accordance with such guidance as the Commission may give to the authority.

(5) Where bodies other than the HSE or local authorities are given enforcement functions they must comply with similar matters as in (4) above.

(6) Nothing in the provisions of this Act or of any regulations made thereunder charging any person in Scotland with the enforcement of any of the relevant statutory provisions shall be construed as authorising that person to institute proceedings for any offence.

(6) This subsection applies to Scotland only and unlike prosecutions In England & Wales, HSE cannot proceed with prosecutions without the sanction of the Procurator Fiscal.

(7) In this Part–

(a) "enforcing authority" means the Executive or any other authority which is by any of the relevant statutory provisions or by regulations under subsection (2) above made responsible for the enforcement of any of those provisions to any extent; and

(b) any reference to an enforcing authority's field of responsibility is a reference to the field over which that authority's responsibility for the enforcement of those provisions extends for the time being;

but where by virtue of paragraph (a) of section 13(1) the performance of any function of the Commission or the Executive is delegated to a government department or person, references to the Commission or the Executive (or to an enforcing authority where that authority is the Executive) in any provision of this Part which relates to that function shall, so far as may be necessary to give effect to any agreement under that paragraph, be construed as references to that department or person; and accordingly any reference to the field of responsibility of an enforcing authority shall be construed as a reference to the field over which that department or person for the time being performs such a function.

> *(7) 'enforcing authority' means the HSE or any other enforcement authority given powers under s18(2) (i.e. local authorities). Any reference to their 'field of responsibility' is a reference to that authority's enforcement responsibilities for the time being.*
>
> *However where HSC or HSE functions are delegated to government departments or persons references to the HSC or HSE shall be taken as reference to that government dept or person who for the time being performs that function.*
>
> *NB: The Environment Act 1995 creates the Environment Agency (England and Wales) & the Scottish Environment Agency (Scotland). Their functions will overlap somewhat with those of HSC/HSE.*

19 Appointment of inspectors

(1) Every enforcing authority may appoint as inspectors (under whatever title it may from time to time determine) such persons having suitable qualifications as it thinks necessary for carrying into effect the relevant statutory provisions within its field of responsibility, and may terminate any appointment made under this section.

> **Section 19 Appointment of Inspectors.**
>
> *(1) HSE, local authorities etc with enforcement powers may appoint inspectors (with total discretion as to job title) from persons having suitable qualifications as they think necessary for enforcement of the Act and all Regulations under the Act. They may also terminate appointments.*

(2) Every appointment of a person as an inspector under this section shall be made by an instrument in writing specifying which of the powers conferred on inspectors by the relevant statutory provisions are to be exercisable by the person appointed; and an inspector shall in right of his appointment under this section—

(a) be entitled to exercise only such of those powers as are so specified; and

(b) be entitled to exercise the powers so specified only within the field of responsibility of the authority which appointed him.

(2) Every appointment of inspectors must be done by a document in writing specifying which powers conferred by s20 or other health and safety regulations the inspector is to have. The inspector must act within the powers (intra vires) granted and not outside his powers (ultra vires) which is illegal. Further, inspectors may be restricted by the field of responsibility of the authority which appointed him (e.g. local authority inspectors would NOT have the power to inspect the nuclear industry)

(3) So much of an inspector's instrument of appointment as specifies the powers which he is entitled to exercise may be varied by the enforcing authority which appointed him.

(3) The appointing authority may vary an inspector's powers in the written document.

(4) An inspector shall, if so required when exercising or seeking to exercise any power conferred on him by any of the relevant statutory provisions, produce his instrument of appointment or a duly authenticated copy thereof.

(4) An inspector may be required to produce his document (or an authenticated copy) indicating which powers he has. Businesses are therefore entitled to ask to see this document when the inspector is purporting to exercise his powers so that confirmation (or otherwise) may be obtained that he is acting within his powers. If an inspector acts outside the powers granted, a case for judicial review of any action taken may be brought.

20   Powers of inspectors

(1) Subject to the provisions of section 19 and this section, an inspector may, for the purpose of carrying into effect any of the relevant statutory provisions within the field of responsibility of the enforcing authority which appointed him, exercise the powers set out in subsection (2) below.

Section 20 – Power of Inspectors

(1) Subject to s19 and 20, inspectors have wide powers to enable compliance with the Act and regulations made under it. These powers are considerably wider in some cases than those possessed by the police and other officials.

(2) The powers of an inspector referred to in the preceding subsection are the following, namely–

(a) at any reasonable time (or, in a situation which in his opinion is or may be dangerous, at any time) to enter any premises which he has reason to believe it is necessary for him to enter for the purpose mentioned in subsection (1) above;

(b) to take with him a constable if he has reasonable cause to apprehend any serious obstruction in the execution of his duty;

(c) without prejudice to the preceding paragraph, on entering any premises by virtue of paragraph (a) above to take with him—

　　(i) any other person duly authorised by his (the inspector's) enforcing authority; and

　　(ii) any equipment or materials required for any purpose for which the power of entry is being exercised;

(d) to make such examination and investigation as may in any circumstances be necessary for the purpose mentioned in subsection (1) above;

(e) as regards any premises which he has power to enter, to direct that those premises or any part of them, or anything therein, shall be left undisturbed (whether generally or in particular respects) for so long as is reasonably necessary for the purpose of any examination or investigation under paragraph (d) above;

(f) to take such measurements and photographs and make such recordings as he considers necessary for the purpose of any examination or investigation under paragraph (d) above;

(g) to take samples of any articles or substances found in any premises which he has power to enter, and of the atmosphere in or in the vicinity of any such premises;

(h) in the case of any article or substance found in any premises which he has power to enter, being an article or substance which appears to him to have caused or to be likely to cause danger to health or safety, to cause it to be dismantled or subjected to any process or test (but not so as to damage or destroy it unless this is in the circumstances necessary for the purpose mentioned in subsection (1) above);

(i) in the case of any such article or substance as is mentioned in the preceding paragraph, to take possession of it and detain it for so long as is necessary for all or any of the following purposes, namely—

　　(i) to examine it and do to it anything which he has power to do under that paragraph;

　　(ii) to ensure that it is not tampered with before his examination of it is completed;

(iii) to ensure that it is available for use as evidence in any proceedings for an offence under any of the relevant statutory provisions or any proceedings relating to a notice under section 21 or 22;

(j) to require any person whom he has reasonable cause to believe to be able to give any information relevant to any examination or investigation under paragraph (d) above to answer (in the absence of persons other than a person nominated by him to be present and any persons whom the inspector may allow to be present) such questions as the inspector thinks fit to ask and to sign a declaration of the truth of his answers;

(k) to require the production of, inspect, and take copies of or of any entry in–

(i) any books or documents which by virtue of any of the relevant statutory provisions are required to be kept; and

(ii) any other books or documents which it is necessary for him to see for the purposes of any examination or investigation under paragraph (d) above;

(l) to require any person to afford him such facilities and assistance with respect to any matters or things within that person's control or in relation to which that person has responsibilities as are necessary to enable the inspector to exercise any of the powers conferred on him by this section;

(m) any other power which is necessary for the purpose mentioned in subsection (1) above.

(2) Inspectors powers are:

(a) to enter any premises during reasonable hours (e.g. business hours) he feels it necessary to enter for the purpose of enforcing the Act any any regulations under the act. He may enter at any time of the day or night if he feels there is a dangerous situation.

(b) to take with him a police officer if he has reasonable fear that he might be obstructed in the exercise of his duty.

(c) in addition to a police officer to take other persons(s) with him provided they are authorised by HSE or LA and any equipment or materials for any purpose for which the power of entry is being exercised (e.g. testing equipment).

(d) to conduct such examinations and investigations as may be necessary to enforce the Act or any regulations under the Act.

(e) regarding premises he has the power to enter to direct that nothing shall be touched or moved so as to preserve evidence and assist the examination or investigation.

(f) to take any measurements, photographs, recordings he considers necessary for the purposes of examination and investigation.

(g) to take sample of any articles or substances found in the premises which he has power to enter to take atmospheric tests in or in the vicinity of those premises.

There has been case law under this section & in some cases evidence has been regarded as inadmissible where inspectors failed to comply strictly to their conditions.

(h) if any article or substances found in the premises has caused danger to health or safety, or is likely to, to order that it be dismantled or subjected to processes and tests but not so as to destroy it unless necessary to enforce the Act or regulations under the Act.

(i) Regarding articles and substances referred to in (h) to take possession and detain them for as long as is necessary to examine it, ensure it is not tampered with prior to complete examination and to produce as evidence in court for alleged offences under the Act/regulations or relating to improvement & prohibition notices (see ss21 & 22).

(j) to require anyone who he reasonably believes may be able to give information relevant to examination/investigation under (d) above to answer such questions as he may think fit and to require a statutory declaration of such statements. The inspector may allow this information to be given alone or in the company of others he has nominated or has approved to be present.

(k) to require the production of and take copies of the whole or any part of:

> *(i) any books or documents required to be kept by the Acts regulations; and*

> *(ii) any other books, documents etc. which it is necessary for him to see to conduct an examination/investigation under (d) above.*

(l) to require anyone to provide him with facilities and assistance regarding matters within that person's control or responsibilities to enable the inspector to carry out his s20 powers. This could extend to office space, telephones, help with manual handling etc. etc.

(m) any other power which is necessary for the enforcement of the Act or regulations under the Act.

(3) The Secretary of State may by regulations make provision as to the procedure to be followed in connection with the taking of samples under subsection (2)(g) above (including provision as to the way in which samples that have been so taken are to be dealt with).

s20(3) The SOS may make regulations regarding procedures to be followed for taking samples under s20(2)(g) including the way in which the samples are to be dealt with.

No regulations have been made as yet.

(4) Where an inspector proposes to exercise the power conferred by subsection (2)(h) above in the case of an article or substance found in any premises, he shall, if so requested by a person who at the time is present in and has responsibilities in

relation to those premises, cause anything which is to be done by virtue of that power to be done in the presence of that person unless the inspector considers that its being done in that person's presence would be prejudicial to the safety of the State.

> *s20(4) Where an inspector proposes to exercise s20(2)(h) powers (see above) he shall, if requested to do so by a person who is present and has responsibilities relating to the premises, exercise those powers in the presence of that person unless the inspector feels that his presence would be prejudicial to the safety of the State.*

(5) Before exercising the power conferred by subsection (2)(h) above in the case of any article or substance, an inspector shall consult such persons as appear to him appropriate for the purpose of ascertaining what dangers, if any, there may be in doing anything which he proposes to do under that power.

> *s20(5) Before exercising s20(2)(h) powers the inspector must consult a appropriate persons for the purpose of ascertaining any dangers he may encounter in the exercise of his powers.*

(6) Where under the power conferred by subsection (2)(i) above an inspector takes possession of any article or substance found in any premises, he shall leave there, either with a responsible person or, if that is impracticable, fixed in a conspicuous position, a notice giving particulars of that article or substance sufficient to identify it and stating that he has taken possession of it under that power; and before taking possession of any such substance under that power an inspector shall, if it is practicable for him to do so, take a sample thereof and give to a responsible person at the premises a portion of the sample marked in a manner sufficient to identify it.

> *s20(6) Where an inspector is executing s20(2)(i) powers he must leave with a responsible person or if that is impracticable put up a prominent notice giving sufficient particulars of the article or substance taken in order to identify it and stating that he has taken possession under s20(2)(i). Before taking possession the inspector must give a marked sample to a responsible person if practicable to do so. This is so that the employer can conduct their own tests and gather evidence to defend themselves.*
>
> *Any seizure in breach of these provisions would be ultra vires and may make matters inadmissible in evidence by the inspector.*

(7) No answer given by a person in pursuance of a requirement imposed under subsection (2)(j) above shall be admissible in evidence against that person or the husband or wife of that person in any proceedings.

> *s20(7) If answers are given by persons under s20(2)(j) they are not admissible in evidence against that person or the spouse of such person in any court proceedings.*

(8) Nothing in this section shall be taken to compel the production by any person of a document of which he would on grounds of legal professional privilege be entitled to withhold production on an order for discovery in an action in the High Court or, as the case may be, on an order for the production of documents in an action in the Court of Session.

> *s20(8) No power in s20 shall compel production of a document which a person would be entitled to withhold by virtue of legal professional privilege even on an order for discovery/production of documents in the High Court (England and Wales) and the Court of Session (Scotland). This protects private communications between lawyers and clients.*
>
> *Definitions:*
>
> *'article' – see s53(1)*
>
> *'substance' – see s53(1)*
>
> *'premises' – see s53(1)*
>
> *'enforcing authority' – see s18(7)(a)*
>
> *'field of responsibility' – see s18(7)(b).*

21 Improvement notices

If an inspector is of the opinion that a person–

    (a) is contravening one or more of the relevant statutory provisions; or

    (b) has contravened one or more of those provisions in circumstances that make it likely that the contravention will continue or be repeated,

he may serve on him a notice (in this Part referred to as "an improvement notice") stating that he is of that opinion, specifying the provision or provisions as to which he is of that opinion, giving particulars of the reasons why he is of that opinion, and requiring that person to remedy the contravention or, as the case may be, the matters occasioning it within such period (ending not earlier than the period within which an appeal against the notice can be brought under section 24) as may be specified in the notice.

> **Section 21 – Improvement Notices**
>
> *An inspector has the power to issue an improvement notice if he is of the opinion that the Act or any regulations under it have been broken or any such contravention may be repeated or continued.*

The notice must specify;

(i) the inspectors opinion as to breach and reasons

(ii) requiring remedial action to be taken

(iii) time scales for compliance for at least 21 days (NB appeals against notice can be brought within 21 days – see s24).

Standard form notices are issued by HSE which give particulars of laws alleged to have been broken and more details of remedial measures etc.

22 Prohibition notices

(1) This section applies to any activities which are being or are [likely] to be carried on by or under the control of any person, being activities to or in relation to which any of the relevant statutory provisions apply or will, if the activities are so carried on, apply.

Section 22 – Prohibition Notices

(1) The power to issue these notices applies to activities being carried on (or likely to be carried on) by or under the control of ANY person provided the activities are encompassed in the Act or Regulations made under the Act.

However, negligence may be sufficient as well as risk flowing from breach of statutory duty.

(2) If as regards any activities to which this section applies an inspector is of the opinion that, as carried on or [likely] to be carried on by or under the control of the person in question, the activities involve or, as the case may be, will involve a risk of serious personal injury, the inspector may serve on that person a notice (in this Part referred to as "a prohibition notice").

(2) If (1) is complied with and an inspector is of the opinion that there is a risk of serious personal injury from the activities he may serve a prohibition notice.

The risk need not be imminent.

If an employer appeals against the notice it is for the inspector to prove on balance of probabilities that there was a risk of serious personal injury. If he establishes this it is then for the employer to prove that all reasonably practicable steps were taken to avoid the risk.

(3) A prohibition notice shall–

(a) state that the inspector is of the said opinion;

(b) specify the matters which in his opinion give or, as the case may be, will give rise to the said risk;

(c) where in his opinion any of those matters involves or, as the case may be, will involve a contravention of any of the relevant statutory provisions, state that he is of that opinion, specify the provision or provisions as to which he is of that opinion, and give particulars of the reasons why he is of that opinion; and

(d) direct that the activities to which the notice relates shall not be carried on by or under the control of the person on whom the notice is served unless the matters specified in the notice in pursuance of paragraph (b) above and any associated contraventions of provisions so specified in pursuance of paragraph (c) above have been remedied.

(3) a prohibition notice must contain the following:

(a) state that the inspector has formed the opinion referred to in (2) above;

(b) specify the matters which in his opinion give rise to a serious risk of personal injury;

(c) specify which parts of the Act/regulations have been broken in his opinion and reasons for that conclusion.

(d) direct that the activities cease unless any risks under (b) above or contravention's under (c) above are remedied

[(4) A direction contained in a prohibition notice in pursuance of subsection (3)(d) above shall take effect–

(a) at the end of the period specified in the notice; or

(b) if the notice so declares, immediately.]

(4) A direction under s22(3)(d) above shall take effect at the end of a period specified in the notice (time necessary to perform safety operations must be taken into account) OR immediately if the notice so specifies.

23 Provisions supplementary to ss 21 and 22

(1) In this section "a notice" means an improvement notice or a prohibition notice.

Section 23 – Supplementary provisions to section 21 and 22.
(1) In s23 'notice' means an improvement notice or a prohibition notice,

(2) A notice may (but need not) include directions as to the measures to be taken to remedy any contravention or matter to which the notice relates; and any such directions–

(a) may be framed to any extent by reference to any approved code of practice; and

(b) may be framed so as to afford the person on whom the notice is served a choice between different ways of remedying the contravention or matter.

> *(2) Notices may (but need not) include directions as to remedial measures to be taken to remedy any beach and such measures may refer to ACOPs and various options as to compliance.*

(3) Where any of the relevant statutory provisions applies to a building or any matter connected with a building and an inspector proposes to serve an improvement notice relating to a contravention of that provision in connection with that building or matter, the notice shall not direct any measures to be taken to remedy the contravention of that provision which are more onerous than those necessary to secure conformity with the requirements of any building regulations for the time being in force to which that building or matter would be required to conform if the relevant building were being newly erected unless the provision in question imposes specific requirements more onerous than the requirements of any such building regulations to which the building or matter would be required to conform as aforesaid.

In this subsection "the relevant building", in the case of a building, means that building, and, in the case of a matter connected with a building, means the building with which the matter is connected.

> *(3) Improvement notices, if they relate to buildings, may NOT impose more stringent provisions than those provided by current Building legislation (see Building Act, 1984).*

(4) Before an inspector serves in connection with any premises used or about to be used as a place of work a notice requiring or likely to lead to the taking of measures affecting the means of escape in case of fire with which the premises are or ought to be provided, he shall consult the fire authority.

In this subsection "fire authority" has the meaning assigned by section 43(1) of the Fire Precautions Act 1971.

> *(4) Notices which require measures to be taken in respect of means of escape in case of fire in respect of workplaces must not be issued without prior consultation with the fire authority under the Fire Precautions Act, 1971.*

(5) Where an improvement notice or a prohibition notice which is not to take immediate effect has been served–

   (a) the notice may be withdrawn by an inspector at any time before the end of the period specified therein in pursuance of section 21 or section 22(4) as the case may be; and

(b) the period so specified may be extended or further extended by an inspector at any time when an appeal against the notice is not pending.

(5) Where notices served are NOT to take immediate effect the inspector may:

(a) withdraw the notice at end time before the end of the period specified therein; and

(b) extend (or further extend) the period so specified at any time when an appeal is not pending.

(6) In the application of this section to Scotland–

(a) in subsection (3) for the words from "with the requirements" to "aforesaid" there shall be substituted the words–

"(a) to any provisions of the building standards regulations to which that building or matter would be required to conform if the relevant building were being newly erected; or

(b) where the sheriff, on an appeal to him under section 16 of the Building (Scotland) Act 1959–

(i) against an order under section 10 of that Act requiring the execution of operations necessary to make the building or matter conform to the building standards regulations, or

(ii) against an order under section 11 of that Act requiring the building or matter to conform to a provision of such regulations,

has varied the order, to any provisions of the building standards regulations referred to in paragraph (a) above as affected by the order as so varied,

unless the relevant statutory provision imposes specific requirements more onerous than the requirements of any provisions of building standards regulations as aforesaid or, as the case may be, than the requirements of the order as varied by the sheriff.";

(b) after subsection (5) there shall be inserted the following subsection–
"(5A) In subsection (3) above ".building standards regulations"" has the s

This sub-section merely amends the wording of s23(3) to make it apply to Scottish Law. In Scotland too improvement notices may not impose more stringent provisions than Scots building legilisation provides.

24 Appeal against improvement or prohibition notice

(1) In this section "a notice" means an improvement notice or a prohibition notice.

Section 24 – Appeal against improvement or prohibition notices

(1) 'Notice' in this section refers either to an improvement notice or a prohibition notice.

(2) A person on whom a notice is served may within such period from the date of its service as may be prescribed appeal to an [employment tribunal]; and on such an appeal the tribunal may either cancel or affirm the notice and, if it affirms it, may do so either in its original form or with such modifications as the tribunal may in the circumstances think fit.

(2) A person on whom a notice is served (usually the employer) may within 21 days from the date of service appeal to an Employment Tribunal. The tribunal may cancel, affirm the order in its original form or subject to modifications as they may in all the circumstances think fit.

(3) Where an appeal under this section is brought against a notice within the period allowed under the preceding subsection, then–

   (a) in the case of an improvement notice, the bringing of the appeal shall have the effect of suspending the operation of the notice until the appeal is finally disposed of or, if the appeal is withdrawn, until the withdrawal of the appeal;

   (b) in the case of a prohibition notice, the bringing of the appeal shall have the like effect if, but only if, on the application of the appellant the tribunal so directs (and then only from the giving of the direction).

(3) (a) appeals against improvement notices suspend the operation of the notice until the appeal is dealt with or the appeal is withdrawn completely
(b) appeals against prohibition orders DO NOT suspend the operation of the notice unless the appellant asks the Tribunal to suspend and they agree. Thus, the prohibition notice continues in force unless the Tribunal directs their lie.

(4) One or more assessors may be appointed for the purposes of any proceedings brought before an [employment tribunal] under this section.

(4) One or more assessors may be appointed to help the tribunal with deliberations under s24.

NB. Rules of procedure are set out in delegated legislation. Pamphlets are available from DfEe outlining this procedure.

25 Power to deal with cause of imminent danger

(1) Where, in the case of any article or substance found by him in any premises which he has power to enter, an inspector has reasonable cause to believe that, in the circumstances in which he finds it, the article or substance is a cause of imminent danger of serious personal injury, he may seize it and cause it to be rendered harmless (whether by destruction or otherwise).

Section 25 – Power to deal with cause of imminent danger
(1) In the case of articles or substances found by the inspector on any premises he has power to enter and he has reasonable cause to believe owing to conditions present that the article or substance is a cause of imminent danger of serious personal injury, he may sieze it and cause it to be rendered harmless by destroying or some other solution.

(2) Before there is rendered harmless under this section–

(a) any article that forms part of a batch of similar articles; or

(b) any substance,

the inspector shall, if it is practicable for him to do so, take a sample thereof and give to a responsible person at the premises where the article or substance was found by him a portion of the sample marked in a manner sufficient to identify it.

(2) Before rendering an article (which forms part of a batch of similar articles) or substance harmless, the inspector must take a sample and give it to a responsible person at the premises where the article or substance was found labelled appropriately in order to identify it.

(3) As soon as may be after any article or substance has been seized and rendered harmless under this section, the inspector shall prepare and sign a written report giving particulars of the circumstances in which the article or substance was seized and so dealt with by him, and shall–

(a) give a signed copy of the report to a responsible person at the premises where the article or substance was found by him; and

(b) unless that person is the owner of the article or substance, also serve a signed copy of the report on the owner;

and if, where paragraph (b) above applies, the inspector cannot after reasonable enquiry ascertain the name or address of the owner, the copy may be served on him by giving it to the person to whom a copy was given under the preceding paragraph.

(3) As soon as possible after exercise of these powers, the inspector must compile a written report giving details of the circumstances in which the sample was seized and shall give signed copies to the responsible person on the premises at the time and the owner of the article or substance if different from the responsible person. If the latters name may not be ascertained after reasonable inquiry the owner's notice may be served on the responsible person.

Definitions:
'article' – see s53(1)
'substance' – see s53(1)
'premises' – see s53(1)
'personal injury' – see s53(1)
'Inspector' – see s53(1)

[25A Power of customs officer to detain articles and substances]

[(1) A customs officer may, for the purpose of facilitating the exercise or performance by an enforcing authority or inspector of any of the powers or duties of the authority or inspector under any of the relevant statutory provisions, seize any imported article or imported substance and detain it for not more than two working days.

Section 25A – Power of customs officer to detain articles and substances
(1) A customs officer (CO) may sieze and detain any imported article or substance for not more than 2 days for the purpose of enabling HSE etc. and inspectors to exercise their powers under the Act or regulations made under the Act.

(2) Anything seized and detained under this section shall be dealt with during the period of its detention in such manner as the Commissioners of Customs and Excise may direct.

(2) Anything seized under s25A(1) shall be dealt with during detention in such manner as the Commissioners for Customs & Excise may direct.

(3) In subsection (1) above the reference to two working days is a reference to a period of forty-eight hours calculated from the time when the goods in question are seized but disregarding so much of any period as falls on a Saturday or Sunday or on Christmas Day, Good Friday or a day which is a bank holiday under the Banking and Financial Dealings Act 1971 in the part of Great Britain where the goods are seized.]

(3) In s25A(1) the reference to '2 days' means a period of 48 hours calculated from the time when they were seized but disregarding any period of time falling on a Saturday or Sunday, or Christmas Day, Good Friday or statutory Bank Holidays.

Definitions:
'customs officer' – see s53(1)
'article' – see s53(1)
'substance' see s53(1)
'inspector' – see s53(1).

NB. The above section was added by the Consumer Protection Act, 1987 s36, Sch 3, para 3.

26 Power of enforcing authorities to indemnify their inspectors

Where an action has been brought against an inspector in respect of an act done in the execution or purported execution of any of the relevant statutory provisions and the circumstances are such that he is not legally entitled to require the enforcing authority which appointed him to indemnify him, that authority may, nevertheless, indemnify him against the whole or part of any damages and costs or expenses which he may have been ordered to pay or may have incurred, if the authority is satisfied that he honestly believed that the act complained of was within his powers and that his duty as an inspector required or entitled him to do it.

Section 26 – Power of Enforcing authorities to indemnify their inspectors
Where a civil legal action is brought against an inspector in respect of acts done in the purported exercise of his statutory duties (see ss20-23 and 25) and circumstances are such that he cannot force HSE/LA to indemnify him against such action, HSE/LA may indemnify him for the whole or part of any damages awarded by the court against him if HSE/LA is satisfied that he honestly believed the act complained of was within his powers (i.e. intra vires) and that his duty as an inspector required or entitled him to do so.
Thus, even though the act may be ultra vires (outside his powers) and the case is won by the complainant, HSE/LA may pay the whole or part of the damages if the condition is satisfied.

OBTAINING AND DISCLOSURE OF INFORMATION

27 Obtaining of information by the Commission, the Executive, enforcing authorities etc

(1) For the purpose of obtaining–

(a) any information which the Commission needs for the discharge of its functions; or

(b) any information which an enforcing authority needs for the discharge of the authority's functions,

the Commission may, with the consent of the Secretary of State, serve on any person a notice requiring that person to furnish to the Commission or, as the case may be, to the enforcing authority in question such information about such matters as may be specified in the notice, and to do so in such form and manner and within such time as may be so specified.

In this subsection "consent" includes a general consent extending to cases of any stated description.

Section 27 – Obtaining of information by the Commission, the Executive, enforcing authorities etc.

(1) For the purpose of obtaining any information HSC or HSE etc. requires for the discharge of its statutory functions, the HSC may, with SoS, consent, serve on any person a notice requiring that person to furnish to HSC or HSE etc. such information as may be specified in the notice, in such form and manner and within such time as may be specified. SOS consent can be a general or specific consent.

(2) Nothing in section 9 of the Statistics of Trade Act 1947 (which restricts the disclosure of information obtained under that Act) shall prevent or penalise–

(a) the disclosure by a Minister of the Crown to the Commission or the Executive of information obtained under that Act about any undertaking within the meaning of that Act, being information consisting of the names and addresses of the persons carrying on the undertaking, the nature of the undertaking's activities, the numbers of persons of different descriptions who work in the undertaking, the addresses or places where activities of the undertaking are or were carried on, the nature of the activities carried on there, or the numbers of persons of different descriptions who work or worked in the undertaking there; ...

(b) ...

(2)(a) Despite s9 of the Statistics of Trade Act, 1947 restricting disclosure of information under that Act a Minister of the Crown may disclose to HSC or HSE any information under that Act about any undertaking (as defined in the 1947 Act) consisting of:

- *names & addresses of persons carrying on the undertaking*
- *nature of the undertaking's activities*
- *number of persons of different descriptions who work in the undertaking*
- *addresses or places where the undertaking's activities were or are carried on*
- *nature of the activities carried on there*
- *number of persons of different descriptions who work or worked in the undertaking there...*

(b) applies to Government training agencies who may also release similar information.

(3) In the preceding subsection any reference to a Minister of the Crown, the Commission, [or the Executive] includes respectively a reference to an officer of his or of that body and also, in the case of a reference to the Commission, includes a reference to–

    (a) a person performing any functions of the Commission or the Executive on its behalf by virtue of section 13(1)(a);

    (b) an officer of a body which is so performing any such functions; and

    (c) an adviser appointed in pursuance of section 13(1)(d).

(3) In s27(2) references to 'Ministers of the Crown', HSC, HSE includes a reference to an officer of such bodies and also in relation to HSC includes:
(a) a person performing HSC or HSE functions under s13(1)(a) – see above
(b) an officer of a body performing such functions; and
(c) an advisor appointed under s13(1)(d) – see above.

(4) A person to whom information is disclosed in pursuance of subsection (2) above shall not use the information for a purpose other than a purpose of the Commission or, as the case may be, of the Executive.

(4) Any information disclosed under s27(2) shall NOT be used for purposes other than the purposes of HSC or HSE.

[27A Information communicated by the Commissioners of Customs and Excise]

[(1) If they think it appropriate to do so for the purpose of facilitating the exercise or performance by any person to whom subsection (2) below applies of any of that person's powers or duties under any of the relevant statutory provisions, the Commissioners of Customs and Excise may authorise the disclosure to that person of any information obtained for the purposes of the exercise by the Commissioners of their functions in relation to imports.

Section 27A – Information communicated by the Commissioners of customs and Excise.
(1) IF CCE think it appropriate to facilitate HSE etc. or an inspector's powers or duties they may authorise the disclosure of information obtained by them in relation to their functions regarding imports.

(2)   This subsection applies to an enforcing authority and to an inspector.

(2) s27A(1) applies to HSE etc. and inspectors.

(3)   A disclosure of information made to any person under subsection (1) above shall be made in such manner as may be directed by the Commissioners of Customs and Excise and may be made through such persons acting on behalf of that person as may be so directed.

(3) A disclosure under s27A(1) shall be made in such manner as may be directed by CCE and may be made through their agents also.

(4)   Information may be disclosed to a person under subsection (1) above whether or not the disclosure of the information has been requested by or on behalf of that person.]

(4) CCE may disclose such information whether or not HSE etc. or an inspector has requested it or not.

28   Restrictions on disclosure of information

(1)   In this and the two following subsections–

   (a) "relevant information" means information obtained by a person under section 27(1) or furnished to any person [under section 27A above or] in pursuance of a requirement imposed by any of the relevant statutory provisions; and

   (b) "the recipient", in relation to any relevant information, means the person by whom that information was so obtained or to whom that information was so furnished, as the case may be.

Section 28 – Restrictions on Disclosure of Information.
(1) In s28(1)(2) and (3);
(a) 'relevant information' means information obtained by a person under s27(1) or furnished to any person under s27A above or in pursuance of the Act or any regulations made under the Act.
(b) 'recipient' in relation to 'relevant information' means the person by whom that information was obtained or to whom the information was furnished.

(2) Subject to the following subsection, no relevant information shall be disclosed without the consent of the person by whom it was furnished.

(2) Subject to s28(3) no 'relevant information' shall be disclosed without the consent of the person by whom it was furnished.

(3) The preceding subsection shall not apply to–

(a) disclosure of information to the Commission, the Executive, [the Environment Agency, the Scottish Environment Protection Agency,] a government department or any enforcing authority;

(b) without prejudice to paragraph (a) above, disclosure by the recipient of information to any person for the purpose of any function conferred on the recipient by or under any of the relevant statutory provisions;

(c) without prejudice to paragraph (a) above, disclosure by the recipient of information to–

(i) an officer of a local authority who is authorised by that authority to receive it,

[(ii) an officer ... of a water undertaker, sewerage undertaker, water authority or water development board who is authorised by that ... undertaker, authority or board to receive it,]

(iii) ...

(iv) a constable authorised by a chief officer of police to receive it;

(d) disclosure by the recipient of information in a form calculated to prevent it from being identified as relating to a particular person or case;

(e) disclosure of information for the purposes of any legal proceedings or any investigation or inquiry held by virtue of section 14(2), or for the purposes of a report of any such proceedings or inquiry or of a special report made by virtue of section 14(2).

(3) s28(2) is NOT applicable to:

(a) disclosure of information to HSC, HSE, Environment Agency, the Scottish Environment Agency, a government department or any enforcing authority (e.g. LAs).

(b) disclosure by the 'recipient' of information to any person for the purpose of any function given to the recipient by the Act or regulations made under the Act, without prejudice to s28(3)(a).

(c) disclosure by the recipient of information to–

　(i)　a local authority officer who is authorised to receive it

　(ii)　an officer of a water undertaking, sewerage undertaking, water authority or water development board authorised by that body to receive it, without prejudice to s28(3)(a).

　(iii)　Similar powers may apply to Rivers Authorities rather than 'river purification boards' (now defunct?).

　(iv)　a police constable authorised by Chief Officer of police to receive it.

(d) disclosure by the recipient of information in such a form that the person or case to which it refers is anonymous.

(e) disclosure of information for the purposes of any legal proceedings or investigation or inquiry (see s14(2)) or for report etc. (see s14(2)). This is to preserve the sub judice rule.

(4) In the preceding subsection any reference to the Commission, the Executive, [the Environment Agency, the Scottish Environment Protection Agency,] a government department or an enforcing authority includes respectively a reference to an officer of that body or authority (including, in the case of an enforcing authority, any inspector appointed by it), and also, in the case of a reference to the Commission, includes a reference to–

(a) a person performing any functions of the Commission or the Executive on its behalf by virtue of section 13(1)(a);

(b) an officer of a body which is so performing any such functions; and

(c) an adviser appointed in pursuance of section 13(1)(d).

(4) In s28(3) references to QUANGO's and government bodies named therein include references to officers of those bodies and inspectors in the case of enforcing bodies. Regarding the HSC it also includes references to persons performing HSC functions under s13(1)(a) and their officers and advisors appointed under s13(1)(d).

(5) A person to whom information is disclosed in pursuance of subsection (3) above shall not use the information for a purpose other than–

(a) in a case falling within paragraph (a) of that subsection, a purpose of the Commission or of the Executive or [of the Environment Agency or of the

Scottish Environment Protection Agency or] of the government department in question, or the purposes of the enforcing authority in question in connection with the relevant statutory provisions, as the case may be;

(b) in the case of information given to an officer of a [body which is a local authority, ... , a water undertaker, a sewerage undertaker, a water authority, a river purification board or a water development board, the purposes of the body] in connection with the relevant statutory provisions or any enactment whatsoever relating to public health, public safety or the protection of the environment;

(c) in the case of information given to a constable, the purposes of the police in connection with the relevant statutory provisions or any enactment whatsoever relating to public health, public safety or the safety of the State.

(5) A person to whom information is disclosed under s28(3) shall NOT use that information other than–

(a) in cases covered by s28(3)(a), for a purpose of the bodies named therein or purposes of the enforcing authority in connection with powers an duties under the Act or regulations under the Act.

(b) in cases covered by s28(3)(c)(i) or (ii), for a statutory purposes of the bodies named therein under appropriate legislation.

(c) in cases covered by s28(3)(c)(iv), for purposes of the police under the Act or regulations under the Act or any legislation whatsoever provided it relates to public health, public safety or the safety of the State.

NB. This attempts to ensure that these bodies do NOT abuse the disclosure requirements in using information for purposes other than their statutory powers.
Judicial review is available if these matters are breached.

[(6) References in subsections (3) and (5) above to a local authority include ... a joint authority established by Part IV of the Local Government Act 1985 [and the London Fire and Emergency Planning Authority].]

(6) Reference in s28(3)&(5) to 'local authority' include a joint authority established by Part IV of the Local Government Act, 1985.

(7) A person shall not disclose any information obtained by him as a result of the exercise of any power conferred by section 14(4)(a) or 20 (including, in particular, any information with respect to any trade secret obtained by him in any premises entered by him by virtue of any such power) except–

(a) for the purposes of his functions; or

(b) for the purposes of any legal proceedings or any investigation or inquiry held by virtue of section 14(2) or for the purposes of a report of any such proceedings or inquiry or of a special report made by virtue of section 14(2); or

(c) with the relevant consent.

In this subsection "the relevant consent" means, in the case of information furnished in pursuance of a requirement imposed under section 20, the consent of the person who furnished it, and, in any other case, the consent of a person having responsibilities in relation to the premises where the information was obtained.

(7) A person shall NOT disclose any information in exercise of powers under a14(2) or s20 (including trade secrets) except–
(a) for the purposes of his statutory functions; or
(b) for purposes of legal procedings, investigations or inquiries or reports under s14(2)
(c) with relevant consent of the owner or the person who furnished it or person having responsibilities in relation to the premises where the information was obtained.
This is to prevent disclosure for purposes other than the enforcement etc. of health & safety law.

(8) Notwithstanding anything in the preceding subsection an inspector shall, in circumstances in which it is necessary to do so for the purpose of assisting in keeping persons (or the representatives of persons) employed at any premises adequately informed about matters affecting their health, safety and welfare, give to such persons or their representatives the following descriptions of information, that is to say–

(a) factual information obtained by him as mentioned in that subsection which relates to those premises or anything which was or is therein or was or is being done therein; and

(b) information with respect to any action which he has taken or proposes to take in or in connection with those premises in the performance of his functions;

and, where an inspector does as aforesaid, he shall give the like information to the employer of the first-mentioned persons.

(8) Despite s28(7) an inspector shall where it is necessary in order to assist in keeping persons or their representatives adequately informed about matters affecting their health, safety and welfare, give to those persons the following:
(a) factual information obtained as mentioned in s28(7) which relates to the premises or anything in them or done in them; and
(b) information in respect of any action which he has taken or proposes to take relating to those premises or in performance of his statutory functions.
The same information must be given to the Employer.

NB. This entitles employees, their union safety representatives or representatives of employee safety to certain information.

[(9) Notwithstanding anything in subsection (7) above, a person who has obtained such information as is referred to in that subsection may furnish to a person who appears to him to be likely to be a party to any civil proceedings arising out of any accident, occurrence, situation or other matter, a written statement of relevant facts observed by him in the course of exercising any of the powers referred to in that subsection.]

(9) Despite s28(7), a person who has received such information may furnish a written statement of relevant facts he observed in the course of his duties to a person who appears to be likely to be a party in any civil proceedings arising out of any accident, occurence, situation or other matter.
Thus, HSE (for example) may release information to assist people in civil claims. Note the discretion, however.

[(10) The Broads Authority and every National Park authority shall be deemed to be local authorities for the purposes of this section.]

Sections 29–32 – Special provisions relating to Agriculture
These sections merely gave similar powers to the Minister of Agriculture to those the SOS has in respect of industry generally. These sections are now repealed (ie are no longer law).

PROVISIONS AS TO OFFENCES

33 Offences

(1) It is an offence for a person–

(a) to fail to discharge a duty to which he is subject by virtue of sections 2 to 7;

(b) to contravene section 8 or 9;

(c) to contravene any health and safety regulations … or any requirement or prohibition imposed under any such regulations (including any requirement or prohibition to which he is subject by virtue of the terms of or any condition or restriction attached to any licence, approval, exemption or other authority issued, given or granted under the regulations);

(d) to contravene any requirement imposed by or under regulations under section 14 or intentionally to obstruct any person in the exercise of his powers under that section;

(e) to contravene any requirement imposed by an inspector under section 20 or 25;

(f) to prevent or attempt to prevent any other person from appearing before an inspector or from answering any question to which an inspector may by virtue of section 20(2) require an answer;

(g) to contravene any requirement or prohibition imposed by an improvement notice or a prohibition notice (including any such notice as modified on appeal);

(h) intentionally to obstruct an inspector in the exercise or performance of his powers or duties [or to obstruct a customs officer in the exercise of his powers under section 25A];

(i) to contravene any requirement imposed by a notice under section 27(1);

(j) to use or disclose any information in contravention of section 27(4) or 28;

(k) to make a statement which he knows to be false or recklessly to make a statement which is false where the statement is made–

(i) in purported compliance with a requirement to furnish any information imposed by or under any of the relevant statutory provisions; or

(ii) for the purpose of obtaining the issue of a document under any of the relevant statutory provisions to himself or another person;

(l) intentionally to make a false entry in any register, book, notice or other document required by or under any of the relevant statutory provisions to be kept, served or given or, with intent to deceive, to made use of any such entry which he knows to be false;

(m) with intent to deceive, to ... use a document issued or authorised to be issued under any of the relevant statutory provisions or required for any purpose thereunder or to make or have in his possession a document so closely resembling any such document as to be calculated to deceive;

(n) falsely to pretend to be an inspector;

(o) to fail to comply with an order made by a court under section 42.

[(1A) Subject to any provision made by virtue of section 15(6)(d), a person guilty of an offence under subsection (1)(a) above consisting of failing to discharge a duty to which he is subject by virtue of sections 2 to 6 shall be liable–

(a) on summary conviction, to a fine not exceeding £20,000;

(b) on conviction on indictment, to a fine.]

(2) A person guilty of an offence under paragraph (d), (f), (h) or (n) of subsection (1) above, or of an offence under paragraph (e) of that subsection consisting of contravening a requirement imposed by an inspector under section 20, shall be liable on summary conviction to a fine not exceeding [level 5 on the standard scale].

[(2A) A person guilty of an offence under subsection (1)(g) or (o) above shall be liable–

(a) on summary conviction, to imprisonment for a term not exceeding six months, or a fine not exceeding £20,000, or both;

(b) on conviction on indictment, to imprisonment for a term not exceeding two years, or a fine, or both.]

(3) Subject to any provision made by virtue of section 15(6)(d) [or (e)] or by virtue of paragraph 2(2) of Schedule 3, a person guilty of [an offence under subsection (1) above not falling within subsection (1A), (2) or (2A) above], or of an offence under any of the existing statutory provisions, being an offence for which no other penalty is specified, shall be liable–

(a) on summary conviction, to a fine not exceeding [the prescribed sum];

(b) on conviction on indictment–

(i) breach of s27(1) notices (disclosure of information)   *JPs: Fine n/e prescribed sum*

CC: Unlimited fine

(j) the use or disclosure of information in breach of s27(4) or section 28.

JPs: Fine n/e prescribed amount

CC: Unlimited fine

(k) knowingly making false or reckless statements where the statement is made

(i) in purported compliance with the Act or any regulations under the Act; or

(ii) for the purpose of obtaining a document himself or for others under the Act or any regulations under the Act.   *Jps: fine n/e prescribed amount*

CC: imprisonment n/e 2 years &/or unlimited fine

(l) intentional false entries in statutory registers, books, notices or other documents with intent to deceive, to be made use of knowing of the falsity. Jps: Fine n/e prescribed amount

CC: unlimited fine

(m) having in one's possession forged documents which closely resemble statutory document with intent to deceive.   *JPs: Fine n/e prescribed amount.*

CC: unlimited fine

(n) impersonation of inspectors   *JPS n/e Level 5 fine.*

(o) failure to comply with court orders under s42 (see below).

Jps: imprisonment n/e 6 months &/or fine n/e 20k

CC: imprisonment n/e 2 years &/or unlimited fine.

NB. Level 5 fine – see current Criminal Justice Act.

"Prescribed fines' may be laid down in regulations made under s15(6)(d) or (e) or Schedule 3, paragraph 2(2) of the Act.

Most of the above offences are triable either in the Magistrates Courts or the Crown Court depending on the seriousness of the offence.

Section 33(1A)

33(2)
33(3)   *These sections lay down the punishments (see above)*

(i)   if the offence is one to which this sub-paragraph applies, to imprisonment for a term not exceeding two years, or a fine, or both;

(ii)   if the offence is not one to which the preceding sub-paragraph applies, to a fine.

(1) List of criminal offences:

Breach of :	*section 2 (employers)*	*JPs*	*20k fine max.*	*CC unlimited fine.*
	section 3 (Employers/ self employed)		*20k fine max.*	*CC unlimited fine*
	section 4 – controllers of premises		*20k fine max.*	*CC unlimited fine*
	section 6 – manufacturers, suppliers, importers, designers etc.		*20k fine max.*	*CC unlimited fine.*
	section 7 – employees		*5k fine max.*	*CC unlimited fine.*

(b) breach of section 8 (any person covered by that section)
> *JPS: 5k max CC unlimited fine*
> *section 9 (employers who charge for safety equipment)*
> *JPs 5k max. CC unlimited fine.*

Note: Imprisonment is NOT available for offences under sections 2–9.

(c) breach of any health & safety regulations or matters stated therein, made under s15
> *JPS 5k fine max. CC unlimited fine.*

Note: If there are breaches of several sections/regulations maximum fines can be awarded in respect of each breach.

(d) breach of regulations made under s14 or intentionally to obstruct any person (usually HSE etc.) in exercise of s14 powers. JPS n/e Level 5 fine.

(e) breach of any inspector's requirement under section 20 or 25
> *JPs (breach of s20)-n/e Level 5 fine.*

(f) to prevent or attempt to prevent any person from assisting an inspector with questions under s20(2). JPs n/e Level 5 fine.

(g) breach of any improvement or prohibition notice (including any modifications on appeal) under sections 21-23.
> *JPs: Imprisonment n/e 6 months &/or fine n/e 20K.*
> *CC: Imprisonment n/e 2 years &/or unlimited fine.*

(h) intentionally to obstruct an inspector or customs officer in the exercise of his statutory powers. (see sections 20, 25 and 25A) JPs n/e Level 5 fine.

(4)   Subsection (3)(b)(i) above applies to the following offences–

(a) an offence consisting of contravening any of the relevant statutory provisions by doing otherwise than under the authority of a licence issued by the Executive ... something for the doing of which such a licence is necessary under the relevant statutory provisions;

(b) an offence consisting of contravening a term of or a condition or restriction attached to any such licence as is mentioned in the preceding paragraph;

(c) an offence consisting of acquiring or attempting to acquire, possessing or using an explosive article or substance (within the meaning of any of the relevant statutory provisions) in contravention of any of the relevant statutory provisions;

(d) ...

(e) an offence under subsection (1)(j) above.

Section 33(3) & (4).
The combined effect of these sub sections is that the following offences are punishable in a Crown Court to imprisonment n/e 2 years &/or an unlimited fine:

(a) contravention of license provisions under the Act or any Regulations under the Act
(b) contravention of terms & conditions in such licences
(c) acquiring or attempting to acquire, possessing or using an explosive article or substance as defined in & in contravention of the Act or any regulations made under the Act.
(d) breach of prohibition notice under s33(1)(g)
(e) breach of disclosure provision under s33(1)(j).

(5) – Repealed by Offshore Safety Act, 1992 ss4(5),(6),7(2) and Schedule 2
(6) – Repealed by Forgery and Counterfeiting Act, 1981 s30, Schedule, Part 1

NB. Other offences are dealt with in sections 36 and 37 (see below).
The Court of Appeal has encouraged magistrates to be more stringent in their fining powers.

34 Extension of time for bringing summary proceedings

(1) Where–

(a) a special report on any matter to which section 14 of this Act applies is made by virtue of subsection (2)(a) of that section; or

(b) a report is made by the person holding an inquiry into any such matter by virtue of subsection (2)(b) of that section; or

(c) a coroner's inquest is held touching the death of any person whose death may have been caused by an accident which happened while he was at work or by a disease which he contracted or probably contracted at work or by any accident, act or omission which occurred in connection with the work of any person whatsoever; or

(d) a public inquiry into any death that may have been so caused is held under the Fatal Accidents Inquiry (Scotland) Act 1895 or the Fatal Accidents and Sudden Deaths Inquiry (Scotland) Act 1906;

and it appears from the report or, in a case falling within paragraph (c) or (d) above, from the proceedings at the inquest or inquiry, that any of the relevant statutory provisions was contravened at a time which is material in relation to the subject-matter of the report, inquest or inquiry, summary proceedings against any person liable to be proceeded against in respect of the contravention may be commenced at any time within three months of the making of the report or, in a case falling within paragraph (c) or (d) above, within three months of the conclusion of the inquest or inquiry.

Section 34 – Extension of time for bringing summary proceedings
(1) Where
(a) an HSC special report is made by reference to section 14(2)(a) powers
(b) a report is made following an inquiry under s14(2)(b)
(c) a coroner's inquest regarding the death of a person possibly caused by an accident or disease whilst at work or by any accident, act or omission which occurred in connection with any person's work
(d) a public inquiry into any death (as in (c) held under Scottish legislation relating to Fatal accidents), and it appears from the documentation or from the inquiry that the Act or any health & safety regulations were broken at the time of the accident etc, proceedings in a magistrates court may be commenced at any time within 3 months of the report or conclusion of the inquest/inquiry.

(2)   Where an offence under any of the relevant statutory provisions is committed by reason of a failure to do something at or within a time fixed by or under any of those provisions, the offence shall be deemed to continue until that thing is done.

(2) Where offences are committed under the Act or regulations are committed by an omission to do something at or within a time fixed by that legislation, the offences shall be deemed to be continuing until the omission is overcome. This also gives the authorities (mainly HSE) more time to bring a magistrates court action.

(3)   Summary proceedings for an offence to which this subsection applies may be commenced at any time within six months from the date on which there comes to the knowledge of a responsible enforcing authority evidence sufficient in the opinion of that authority to justify a prosecution for that offence; and for the purposes of this subsection–

(a) a certificate of an enforcing authority stating that such evidence came to its knowledge on a specified date shall be conclusive evidence of that fact; and

(b) a document purporting to be such a certificate and to be signed by or on behalf of the enforcing authority in question shall be presumed to be such a certificate unless the contrary is proved.

(3) Magistrates proceedings under section 34 may be commenced at any time within 6 months from the day on which it came to HSE etc's knowledge that there is sufficient evidence to justify a prosecution & for these purposes—
(a) an HSE etc. certificate stating such evidence came to light on a specified date shall be unchallengable evidence of that fact; and
(b) a document purporting to be a document as in (a) shall be deemed to be so unless the contrary is proved.
This is to avoid procedural disputes and wrangles in court.
NB. This subsection only applies to manufacturers etc. (see s34(5) below).

(4) The preceding subsection applies to any offence under any of the relevant statutory provisions which a person commits by virtue of any provision or requirement to which he is subject as the designer, manufacturer, importer or supplier of any thing; and in that subsection "responsible enforcing authority" means an enforcing authority within whose field of responsibility the offence in question lies, whether by virtue of section 35 or otherwise.

(4) Section (3) applies to any offence under the Act (e.g. s6) or regulations applicable to designers, manufacturers, importers or suppliers of anything provided the enforcing authority has the appropriate field of responsibility under section 35 or by other statutory provisions (e.g. s18(7)(b)).

(5) In the application of subsection (3) above to Scotland—

(a) for the words from "there comes" to "that offence" there shall be substituted the words "evidence, sufficient in the opinion of the enforcing authority to justify a report to the Lord Advocate with a view to consideration of the question of prosecution, comes to the knowledge of the authority";

(b) at the end of paragraph (b) there shall be added the words

"and

(c) [section 331(3) of the Criminal Procedure (Scotland) Act 1975] (date of commencement of proceedings) shall have effect as it has effect for the purposes of that section."

[(6) In the application of subsection (4) above to Scotland, after the words "applies to" there shall be inserted the words "any offence under section 33(1)(c) above where the health and safety regulations concerned were made for the general purpose mentioned in section 18(1) of the Gas Act 1986 and".]

35 Venue

An offence under any of the relevant statutory provisions committed in connection with any plant or substance may, if necessary for the purpose of bringing the offence within the field of responsibility of any enforcing authority or conferring jurisdiction on any court to entertain proceedings for the offence, be treated as having been committed at the place where that plant or substance is for the time being.

> ## Section 35 – Venue
> *Criminal prosecutions under the Act or the regulations relating to any plant or substance, may be brought by any HSE office etc. and in any court where the plant or substance is for the time being even if the offence was committed elsewhere. This gives HSE etc. flexibility regarding court venue.*
>
> *Definitions: 'plant', 'substance' – see s53(1)*

36 Offences due to fault of other person

(1) Where the commission by any person of an offence under any of the relevant statutory provisions is due to the act or default of some other person, that other person shall be guilty of the offence, and a person may be charged with and convicted of the offence by virtue of this subsection whether or not proceedings are taken against the first-mentioned person.

> ## Section 36 – Offences due to fault of other person
> *(1) Where a criminal offence under the Act or the regulations is committed by any person (X) but the act is due to the act or default of some other person (Y), that other person (Y) shall be guilty of the offence & may be charged and convicted whether or not proceedings are taken against the first person (X).*
> *Thus, it may come to light in the course of prosecution of an employer that the action was the fault of an employee (especially employees entrusted as 'competent persons for various purposes, eg diving operations)). HSE can prosecute BOTH or choose who to prosecute. NB. In a lot of cases there will be breaches by both employer (under s2) and employee (under s7) & HSE can proceed against both if they wish or, as is more usual employer only. Other people can also be prosecuted in addition to or instead of the employer, i.e. Self employed (s3), those in control of premises (s4), manufacturers etc. (s6), employees (s7) any person (s8) and directors etc. (see s37).*

(2) Where there would be or have been the commission of an offence under section 33 by the Crown but for the circumstance that that section does not bind the Crown, and that fact is due to the act or default of a person other than the Crown, that person shall be guilty of the offence which, but for that circumstance, the Crown would be committing or would have committed, and may be charged with and convicted of that offence accordingly.

(2) Although the Crown cannot be prosecuted for criminal offences by virtue of the mediaeval maxim "The Queen can do no wrong" and this includes Departments of State etc, nevertheless if offences are committed on behalf of Crown bodies the individual civil servant etc. responsible may be prosecuted but NOT the Crown.

(3)   The preceding provisions of this section are subject to any provision made by virtue of section 15(6).

(3) Sections 36(1) and (2) are subject to any regulations made under s15(6) (see above)

37   Offences by bodies corporate

(1)   Where an offence under any of the relevant statutory provisions committed by a body corporate is proved to have been committed with the consent or connivance of, or to have been attributable to any neglect on the part of, any director, manager, secretary or other similar officer of the body corporate or a person who was purporting to act in any such capacity, he as well as the body corporate shall be guilty of that offence and shall be liable to be proceeded against and punished accordingly.

Section 37 – Offences by bodies corporate
(1) Where criminal offences under the Act or the regulations have been committed by a Corporation (i.e. Companies under the Companies Act, Statutory Corporations (including QUANGO's, local authorities, statutory undertakers etc., Common Law Corporations, etc) that Corporation may be fined. However, if the criminal acts are proved to have been committed with the consent or contrivance or neglect of any director, manager, secretary or other similar officer acting in that capacity, he as well as the Corporation may be found guilty of the offence and punished.
Thus, although these higher paid managerial people are employees they may be held accountable personally for breaches they helped to set up. The reason for this is that a Corporation (although a person in law) has to act through its human agents. These human agents are usually higher paid staff and it is felt right that they should have some responsibility.
NB. Several cases have been brought against directors under this section. In addition some directors have also been prosecuted for manslaughter at Common Law. A further punishment which has been meted out to some directors is disqualification for a period of time under the Disqualification of Directors legislation.

(2)   Where the affairs of a body corporate are managed by its members, the preceding subsection shall apply in relation to the acts and defaults of a member in

connection with his functions of management as if he were a director of the body corporate.

> *(2) Where Corporations are managed by its members (e.g. certain co-operatives and common ownership companies) action may be taken against any of those managing members under s37(1) as if he were a director etc.*
>
> *NB. Definitions of 'attributable to', 'neglect' 'manager' 'similar' 'officer' are contained in series of decided cases as the Act does not define these terms.*

38   Restriction on institution of proceedings in England and Wales

Proceedings for an offence under any of the relevant statutory provisions shall not, in England and Wales, be instituted except by an inspector or [the Environment Agency or] by or with the consent of the Director of Public Prosecutions.

> **Section 38 – Restriction on Institution of proceedings in England and Wales**
> *Criminal proceedings for offences under the Act or the regulations in England & Wales can only be commenced by an HSE/LA inspector or the Environment Agency by or with the consent of the Director of Public Prosecutions (DPP).*
> *NB. Crown Prosecution Service (CPS) will decide if manslaughter actions will be brought.*

39   Prosecutions by inspectors

(1)   An inspector, if authorised in that behalf by the enforcing authority which appointed him, may, although not of counsel or a solicitor, prosecute before a magistrates' court proceedings for an offence under any of the relevant statutory provisions.

> **Section 39 – Prosecution by inspectors**
> *(1) Inspectors, if authorised by HSE etc, may bring magistrates court proceedings for offences under the Act or regulations even though they are not barristers or solicitors.*

(2)   This section shall not apply to Scotland.

> *(2) This does not apply to Scotland.*

40   Onus of proving limits of what is practicable etc.

In any proceedings for an offence under any of the relevant statutory provisions consisting of a failure to comply with a duty or requirement to do something so far as is practicable or so far as is reasonably practicable, or to use the best means to do something, it shall be for the accused to prove (as the case may be) that it was not practicable or not reasonably practicable to do more than was in fact done to satisfy the duty or requirement, or that there was no better practicable means than was in fact used to satisfy the duty or requirement.

> **Section 40 – Onus of proving limits of what is practicable etc.**
> *In any criminal proceedings under the Act or regulations where the charge involves failure to comply with a statutory duty:*
> *(a) 'so far as is practicable'; or*
> *(b) 'so far as is reasonably practicable' (e.g. see s2); or*
> *(c) to 'use best practicable means',*
> *it is for the defendant to prove that he satisfied these standards by establishing what he actually did complied with (a)(b) or (c) as appropriate.*
> *Thus, HSE inspectors etc. merely have to bring the charge. The burden of proving compliance is with the defendant. Howewver, this burden of proof established 'on a balance of probabilities' and NOT 'beyond all reasonable doubt.'*

41 Evidence

(1) Where an entry is required by any of the relevant statutory provisions to be made in any register or other record, the entry, if made, shall, as against the person by or on whose behalf it was made, be admissible as evidence or in Scotland sufficient evidence of the facts stated therein.

> **Section 41 – Evidence**
> *(1) Where the Act or any regulations require entries in registers, records etc, those entries (if made) shall be admissible as evidence against the person by or on behalf or whom it was made. In Scotland it shall be 'sufficient evidence' of the facts stated therein.*

(2) Where an entry which is so required to be so made with respect to the observance of any of the relevant statutory provisions has not been made, that fact shall be admissible as evidence or in Scotland sufficient evidence that that provision has not been observed.

> *(2) Where an entry has NOT been made that also is admissible in evidence ('sufficient evidence in Scotland') that the Act or regulations have not been complied with.*

42 Power of court to order cause of offence to be remedied or, in certain cases, forfeiture

(1) Where a person is convicted of an offence under any of the relevant statutory provisions in respect of any matters which appear to the court to be matters which it is in his power to remedy, the court may, in addition to or instead of

imposing any punishment, order him, within such time as may be fixed by the order, to take such steps as may be specified in the order for remedying the said matters.

Section 42 – Power of court to order cause of offence to be remedied or, in certain cases, forfeiture
(1) Where convictions are secured under the Act or regulations and the court feels the person convicted may remedy the situation, the court may order that remedy in addition to or instead of other punishments and within the timescale specified in the court order.

(2) The time fixed by an order under subsection (1) above may be extended or further extended by order of the court on an application made before the end of that time as originally fixed or as extended under this subsection, as the case may be.

(2) The time fixed in s42(1) may be extended or further extended by court order provided an application is made to the court before the expiry of the previous order.

(3) Where a person is ordered under subsection (1) above to remedy any matters, that person shall not be liable under any of the relevant statutory provisions in respect of those matters in so far as they continue during the time fixed by the order or any further time allowed under subsection (2) above.

(3) Where under s42(1) a person is ordered to remedy a situation, he shall NOT be liable to additional penalties under the Act or regulations during the continuance of the time fixed for compliance (or extension of time).

(4) Subject to the following subsection, the court by or before which a person is convicted of an offence such as is mentioned in section 33(4)(c) in respect of any such explosive article or substance as is there mentioned may order the article or substance in question to be forfeited and either destroyed or dealt with in such other manner as the court may order.

(4) If a person is convicted of an offence (such as a s33(4)(c) offence) in respect of explosive articles or substances, the court may order forfeiture and destruction or such other dealing as the court may order, subject to s42(5) below.

(5) The court shall not order anything to be forfeited under the preceding subsection where a person claiming to be the owner of or otherwise interested in it applies to

be heard by the court, unless an opportunity has been given to him to show cause why the order should not be made.

> *(5) Forfeiture powers (but NOT destruction powers, etc) under s42(4) shall NOT be exercised where a person claiming to be the owner or having other property interest in the subject matter applies to be heard by the court, unless an opportunity has already been given to him to plead why the order should not be made.*

FINANCIAL PROVISIONS

43 Financial provisions

(1) It shall be the duty of the Secretary of State to pay to the Commission such sums as are approved by the Treasury and as he considers appropriate for the purpose of enabling the Commission to perform its functions; and it shall be the duty of the Commission to pay to the Executive such sums as the Commission considers appropriate for the purpose of enabling the Executive to perform its functions.

> **Section 43 – Financial Provisions**
> *(1) It is the duty of the SOS to pay to HSC such sums as are approved by the Treasury as he considers appropriate to enable HSC to perform its statutory functions.*
> *HSC has a duty to pay HSE such sums as HSC considers appropriate to enable HSE to perform its statutory functions.*
>
> *NB. These sums have never been enough and HSC and HSE are under-resourced and understaffed which, of course, minimises their effective enforcement functions.*

(2) Regulations may provide for such fees as may be fixed by or determined under the regulations to be payable for or in connection with the performance by or on behalf of any authority to which this subsection applies of any function conferred on that authority by or under any of the relevant statutory provisions.

> *(2) Regulations may provide for fees to be charged by, SoS, HSC, HSE, LA's etc. in respect of any of their functions.*
> *NB. Regulations do allow for fees to be charged to employers by EMAS for medical examinations.*

(3) Subsection (2) above applies to the following authorities, namely the Commission, the Executive, the Secretary of State, ... every enforcing authority, and any other person on whom any function is conferred by or under any of the relevant statutory provisions.

(3) This merely spells out the authorities who may charge (see s43(2) above).

(4) Regulations under this section may specify the person by whom any fee payable under the regulations is to be paid; but no such fee shall be made payable by a person in any of the following capacities, namely an employee, a person seeking employment, a person training for employment, and a person seeking training for employment.

(4) Regulations under s43(2) may specify who is to be charged (i.e. mainly employers) but no fee shall be payable by:
(a) employees
(b) job seekers
(c) trainees
(d) persons seeking training for employment.

(5) Without prejudice to section 82 (3), regulations under this section may fix or provide for the determination of different fees in relation to different functions, or in relation to the same function in different circumstances.

(5) Regulations under s43(2) may specify different fees for different functions or the same function in different circumstances. This is without prejudice to s82(3) powers (see below).

[(6) The power to make regulations under this section shall be exercisable by the Secretary of State, the Minister of Agriculture, Fisheries and Food or the Secretary of State and that Minister acting jointly.]

(6) The power to make regulations under s43(2) belongs to SoS, MAFF or both of them acting jointly.

(7) repealed.

(8) In subsection (4) above the references to a person training for employment and a person seeking training for employment shall include respectively a person attending an industrial rehabilitation course provided by virtue of the Employment and Training Act 1973 and a person seeking to attend such a course.

(8) In s43(4) references to a 'person training for employment' and 'a person seeking training for employment' shall include respectively a person attending an industrial rehabilitation course under the Employment & Training Act, 1973 and a person seeking to attend such a course.

(9) For the purposes of this section the performance by an inspector of his functions shall be treated as the performance by the enforcing authority which appointed him of functions conferred on that authority by or under any of the relevant statutory provisions.

(9) For s43 purposes, the performance by an inspector of his statutory functions shall be treated as performance of statutory functions by HSE, etc. (i.e. the inspector is an agent of HSE etc.; he does not act in a personal capacity).

MISCELLANEOUS AND SUPPLEMENTARY

44 Appeals in connection with licensing provisions in the relevant statutory provisions

(1) Any person who is aggrieved by a decision of an authority having power to issue licences (other than ... nuclear site licences) under any of the relevant statutory provisions–

(a) refusing to issue him a licence, to renew a licence held by him, or to transfer to him a licence held by another;

(b) issuing him a licence on or subject to any term, condition or restriction whereby he is aggrieved;

(c) varying or refusing to vary any term, condition or restriction on or subject to which a licence is held by him; or

(d) revoking a licence held by him,

may appeal to the Secretary of State.

Section 44 – Appeals in connection with licensing provisions in the relevant statutory provisions

(1) Anyone who is aggrieved by a decison of a body having power to issue licences under the Act or regulations (except nuclear site licences) may appeal to SoS against–

(a) refusal to issue or renew a licence or to transfer a licence held by someone else;

(b) terms, conditions or restrictions imposed in the issue of a licence;

(c) refusal to vary any term, condition or restriction on or subject to which a licence is held; or

(d) revocation of a licence held by him.

(2) The Secretary of State may, in such cases as he considers it appropriate to do so, having regard to the nature of the questions which appear to him to arise, direct that an appeal under this section shall be determined on his behalf by a person appointed by him for that purpose.

(2) SoS may deputise others for appeal purposes.

(3) Before the determination of an appeal the Secretary of State shall ask the appellant and the authority against whose decision the appeal is brought whether they wish to appear and be heard on the appeal and–

(a) the appeal may be determined without a hearing of the parties if both of them express a wish not to appear and be heard as aforesaid;

(b) the Secretary of State shall, if either of the parties expresses a wish to appear and be heard, afford to both of them an opportunity of so doing.

(3) Before determining appeals the SoS shall ask the appellant and respondent whether they wish to appear and be heard &–

(a) the appeal may go ahead in the absence of the parties if both agree

(b) SoS if either party wishes to appear may afford both parties an opportunity of

(4) The Tribunals and Inquiries Act [1992] shall apply to a hearing held by a person appointed in pursuance of subsection (2) above to determine an appeal as it applies to a statutory inquiry held by the Secretary of State, but as if in [section 10(1)] of that Act (statement of reasons for decisions) the reference to any decision taken by the Secretary of State included a reference to a decision taken on his behalf by that person.

(4) Procedures for such appeals will generally follow the same procedures as laid down in the Tribunals & Inquiries Act, 1992 subject to the addition of decisions being reached by deputies appointed under s44(2).

(5) A person who determines an appeal under this section on behalf of the Secretary of State and the Secretary of State, if he determines such an appeal, may give such directions as he considers appropriate to give effect to his determination.

> *(5) The SoS (or his deputy) in determining an appeal may give such directions as he thinks appropriate to give effect to his decision.*

(6) The Secretary of State may pay to any person appointed to hear or determine an appeal under this section on his behalf such remuneration and allowances as the Secretary of State may with the approval of the Minister for the Civil Service determine.

> *(6) SoS may pay deputies appointed under s44(2) remuneration and expenses subject to approval of the Minister for the Civil Service (now the Treasury).*

(7) In this section–

   (a) "licence" means a licence under any of the relevant statutory provisions other than [a nuclear site licence];

   (b) "nuclear site licence" means a licence to use a site for the purpose of installing or operating a nuclear installation within the meaning of the following subsection.

> *(7) (a) 'licence' means a licence under the Act or any regulations (but not a nuclear licence)*
> *(b) 'nuclear site licence' means a licence to use a site for the purpose of installing or operating a nuclear installation as defined in s44(8) below.*

(8) For the purposes of the preceding subsection "nuclear installation" means–

   (a) a nuclear reactor (other than such a reactor comprised in a means of transport, whether by land, water or air); or

   (b) any other installation of such class or description as may be prescribed for the purposes of this paragraph or section 1(1)(b) of the Nuclear Installations Act 1965, being an installation designed or adapted for–

   (i) the production or use of atomic energy; or

   (ii) the carrying out of any process which is preparatory or ancillary to the production or use of atomic energy and which involves or is capable of causing the emission of ionising radiations; or

   (iii) the storage, processing or disposal of nuclear fuel or of bulk quantities of other radioactive matter, being matter which has been produced or irradiated in the course of the production or use of nuclear fuel;

and in this subsection–

"atomic energy" has the meaning assigned by the Atomic Energy Act 1946;

"nuclear reactor" means any plant (including any machinery, equipment or appliance whether affixed to land or not) designed or adapted for the production of atomic energy by a fission process in which a controlled chain reaction can be maintained without an additional source of neutrons.

(8) In s44(7) 'nuclear installation' means–

(a) a nuclear reactor (other than one comprised in a means of transport by land, water or air); or

(b) any other such installation as may be prescribed for s44(8)(b) purposes or s1(1)(b) Nuclear Installations Act, 1965 being an installation designed or adapted for–

(i) the production or use of nuclear energy; or

(ii) the carrying out of any process preparatory to or ancillary to the production or use of atomic energy & which involves or is capable of emitting ionising radiations; or

(iii) the storage, processing or disposal of nuclear fuel or bulk quantities of other radio active materials which has been produced or irradiated in the course or production or use of nuclear fuel.;

& in s44(8) 'atomic energy' has the meaning assigned it by the Atomic Energy Act 1946; 'nuclear reactor' means any plant (including machinery, equipment or appliance whether a fixture or not) designed or adapted for the production of nuclear energy by a fission process in which a controlled chain reaction can be maintained without an additional source of neutrons.

45 Default powers

(1) Where, in the case of a local authority who are an enforcing authority, the Commission is of the opinion that an investigation should be made as to whether that local authority have failed to perform any of their enforcement functions, the Commission may make a report to the Secretary of State.

Section 45 Default Powers
(1) Where LA's are the enforcing authority and have failed to perform their enforcement functions & HSC are of opinion that an investigation should be made, HSC may report to the SoS.

(2) The Secretary of State may, after considering a report submitted to him under the preceding subsection, cause a local inquiry to be held; and the provisions of subsections (2) to (5) of section 250 of the Local Government Act 1972 as to

local inquiries shall, without prejudice to the generality of subsection (1) of that section, apply to a local inquiry so held as they apply to a local inquiry held in pursuance of that section.

(2) The SoS after considering the s45(1) report, may cause a local inquiry to be held. Procedures under s250(2)–(5) Local Government Act 1972 shall apply without prejudice to the generality of s225(1).

(3) If the Secretary of State is satisfied, after having caused a local inquiry to be held into the matter, that a local authority have failed to perform any of their enforcement functions, he may make an order declaring the authority to be in default.

(3) If, after an inquiry, the SoS is satisfied that the LA have failed to perform their function's he may make a default order.

(4) An order made by virtue of the preceding subsection which declares an authority to be in default may, for the purpose of remedying the default, direct the authority (hereafter in this section referred to as "the defaulting authority") to perform such of their enforcement functions as are specified in the order in such manner as may be so specified and may specify the time or times within which those functions are to be performed by the authority.

(4) A default order may direct the LA to perform specified enforcement functions and time scales.

(5) If the defaulting authority fail to comply with any direction contained in such an order the Secretary of State may, instead of enforcing the order by mandamus, make an order transferring to the Executive such of the enforcement functions of the defaulting authority as he thinks fit.

(5) If the LA fails to comply with the default order directions, SoS may transfer to HSE its functions instead of by mandamus.
NB. Mandamus is a Common Law remedy whereby the High Court can order a statutory body to perform its functions where they are failing to perform them.

(6) Where any enforcement functions of the defaulting authority are transferred in pursuance of the preceding subsection, the amount of any expenses which the Executive certifies were incurred by it in performing those functions shall on demand be paid to it by the defaulting authority.

(6) Where LA powers are transferred to HSE, HSE may charge the LA for such functions & the LA must pay it on demand.

(7) Any expenses which in pursuance of the preceding subsection are required to be paid by the defaulting authority in respect of any enforcement functions transferred in pursuance of this section shall be defrayed by the authority in the like manner, and shall be debited to the like account, as if the enforcement functions had not been transferred and the expenses had been incurred by the authority in performing them.

(7) Any expenses under s45(6) required to be paid by the LA, shall be defrayed by the LA in like manner, & shall be debited to the like account, as if the enforcement functions had NOT been transferred & the LA had incurred the expenses.
NB. This is so that these expenses appear in LA accounts, which are open to audit and public accountability.

(8) Where the defaulting authority are required to defray any such expenses the authority shall have the like powers for the purpose of raising the money for defraying those expenses as they would have had for the purpose of raising money required for defraying expenses incurred for the purpose of the enforcement functions in question.

(8) If an LA lacks the money to defray these expenses it may raise the money (e.g. by an increase in Council Tax) as if it were performing those enforcement functions (i.e. ultimately the taxpayer pays).

(9) An order transferring any enforcement functions of the defaulting authority in pursuance of subsection (5) above may provide for the transfer to the Executive of such of the rights, liabilities and obligations of the authority as the Secretary of State considers appropriate; and where such an order is revoked the Secretary of State may, by the revoking order or a subsequent order, make such provision as he considers appropriate with respect to any rights, liabilities and obligations held by the Executive for the purposes of the transferred enforcement functions.

(9) A transfer order under s45(5) may provide for the transfer to HSE of such rights, liabilities and obligations as the SoS considers appropriate and he may provide in like manner on the revocation of orders or subsequent orders.

(10) The Secretary of State may by order vary or revoke any order previously made by him in pursuance of this section.

(10) SoS may by order vary or revoke any previous order made by him.

(11) In this section "enforcement functions", in relation to a local authority, means the functions of the authority as an enforcing authority.

(11) In s45 'enforcement functions' relating to a LA means the functions of the LA as enforcing authority under the Act or regulations under the Act.

(12) In the application of this section to Scotland–

(a) in subsection (2) for the words " subsections (2) to (5) of section 250 of the Local Government Act 1972" there shall be substituted the words " subsections (2) to (8) of section 210 of the Local Government (Scotland) Act 1973", except that before 16th May 1975 for the said words there shall be substituted the words " subsections (2) to (9) of section 355 of the Local Government (Scotland) Act 1947";

(b) in subsection (5) the words "instead of enforcing the order by mandamus" shall be omitted.

46   Service of notices

(1) Any notice required or authorised by any of the relevant statutory provisions to be served on or given to an inspector may be served or given by delivering it to him or by leaving it at, or sending it by post to, his office.

Section 46 – Service of Notices
(1) Any notice required or authorised under the Act/regulations to be served on or given to an inspector may be served or given by delivering it to him or leaving it at, or sending it by post to his office.
NB. To avoid disputes it is advisable to get somebody to sign a receipt or to send by recorded delivery/registered Post.

(2) Any such notice required or authorised to be served on or given to a person other than an inspector may be served or given by delivering it to him, or by leaving it at his proper address, or by sending it by post to him at that address.

(2) Any notice required or authorised under the Act/regulations to be served on or given to other persons may be served or given by delivering it to him or leaving it at, or sending it by post to his proper address.
NB. To avoid disputes it is advisable to get somebody to sign a receipt or to send by recorded delivery/registered post.

(3) Any such notice may–

(a) in the case of a body corporate, be served on or given to the secretary or clerk of that body;

(b) in the case of a partnership, be served on or given to a partner or a person having the control or management of the partnership business or, in Scotland, the firm.

> *(3) Any notice may–*
>
> *(a) in the case of Companies/Corporations be served on or given to the Company Secretary or Clerk of that organisation;*
>
> *(b) in the case of partnerships, be served on or given to any partner or person having the control or management of such a business or, in Scotland the firm.*

(4) For the purposes of this section and of section 26 of the Interpretation Act 1889 (service of documents by post) in its application to this section, the proper address of any person on or to whom any such notice is to be served or given shall be his last known address, except that–

(a) in the case of a body corporate or their secretary or clerk, it shall be the address of the registered or principal office of that body;

(b) in the case of a partnership or a person having the control or the management of the partnership business, it shall be the principal office of the partnership;

and for the purposes of this subsection the principal office of a company registered outside the United Kingdom or of a partnership carrying on business outside the United Kingdom shall be their principal office within the United Kingdom.

> *(4) For the purposes of s46 & s26 Interpretation Act, 1889 (now 1978) (service of documents by post) as applied to s46, the proper address of any person shall be his last known address, except that–*
>
> *(a) in the case of a Company/Corporation or their Secretary or Clerk, it shall be the registered office or principal office of that organisation;*
>
> *(b) in the case of a partnership or person having control of the business, it shall be the principal office of the partnership.*
>
> *For s46(4) purposes the principal office of a Company registered outside the UK shall be their principal office within the UK.*

(5) If the person to be served with or given any such notice has specified an address within the United Kingdom other than his proper address within the meaning of

subsection (4) above as the one at which he or someone on his behalf will accept notices of the same description as that notice, that address shall also be treated for the purposes of this section and section 26 of the Interpretation Act 1889 as his proper address.

(5) If a person in the UK has specified an address other than his proper address within the UK for service etc, that address shall be treated as valid for s46 and s26 Interpretation Act 1889 (now 1978) purposes.

(6) Without prejudice to any other provision of this section, any such notice required or authorised to be served on or given to the owner or occupier of any premises (whether a body corporate or not) may be served or given by sending it by post to him at those premises, or by addressing it by name to the person on or to whom it is to be served or given and delivering it to some responsible person who is or appears to be resident or employed in the premises.

(6) Without prejudice to s46(1) to 95) and (7)–(8) any notice to be served on the owner or occupier of any premises (whether a Company or not) may be served by sending it by post to him at those premises or by addressing it to his agent at those premises or delivering it to some responsible person who appears to be a resident or employed at those premises.

(7) If the name or the address of any owner or occupier of premises on or to whom any such notice as aforesaid is to be served or given cannot after reasonable inquiry be ascertained, the notice may be served or given by addressing it to the person on or to whom it is to be served or given by the description of "owner" or "occupier" of the premises (describing them) to which the notice relates, and by delivering it to some responsible person who is or appears to be resident or employed in the premises, or, if there is no such person to whom it can be delivered, by affixing it or a copy of it to some conspicuous part of the premises.

(7) If owners or occupiers names cannot be ascertained after reasonable inquiry, notices may be served or given by addressing it to the owner or occupier by name or as 'owner' or 'occupier' of the premises and by delivering it to some responsible person resident or employed at the premises or (if there is no-one who can be personally served) by affixing a prominent notice at the premises.

(8) The preceding provisions of this section shall apply to the sending or giving of a document as they apply to the giving of a notice.

(8) S46(1) to (7) shall apply to the sending or giving of all documents as well as notices.

NB. This whole section is designed to limit disputes about whether a notice was served or not and will, save the parties and the courts time.

Definitions:
'premises'—see s53(1)
'registered office'—see Companies Act,1985 s287.

47 Civil liability

(1) Nothing in this Part shall be construed–

    (a) as conferring a right of action in any civil proceedings in respect of any failure to comply with any duty imposed by sections 2 to 7 or any contravention of section 8; or

    (b) as affecting the extent (if any) to which breach of a duty imposed by any of the existing statutory provisions is actionable; or

    (c) as affecting the operation of section 12 of the Nuclear Installations Act 1965 (right to compensation by virtue of certain provisions of that Act).

Section 47 – Civil Liability
(1) Nothing in Part 1 of this Act (i.e. sections 1 to 54) shall be interpreted
(a) as giving a right to sue at Civil Law in respect of breaches of section 2 to 8; or
(b) as affecting any Civil action available (if any) under any previous legislation (e.g. Factories Act etc.); or
(c) as affecting actions for compensation under Nuclear Installations Act,1965.

This Act is purely a criminal law measure and gives rise to no automatic civil actions for breach of statutory duty. Breach of statutory duty was available under certain older pieces of legislation and these continue until such time as they are repealed. The latter process is almost complete now.

This should not be confused with a right to refer to criminal actions as proof in one's civil actions for negligence, for example. Breach of Statutory duty actions are NOT available.

(2) Breach of a duty imposed by health and safety regulations … shall, so far as it causes damage, be actionable except in so far as the regulations provide otherwise.

> *(2) Breach of statutory duty as laid down in health & safety regulations shall (if damage is caused) be actionable unless the regulations provide otherwise.*
>
> *The MHSWR reg. 22 (see below) do provide otherwise and except in the case of young persons and pregnant women and nursing mothers no action may be brought at civil law for breach of statutory duty.*
>
> *Most other regulations are silent as to civil liability in which case such actions may be brought if the conditions are satisfied (especially damage). Guarding provisions under PUWER 1998 for example, are likely to lead to breach of statutory duty actions at Civil law.*

(3) No provision made by virtue of section 15(6)(b) shall afford a defence in any civil proceedings, whether brought by virtue of subsection (2) above or not; but as regards any duty imposed as mentioned in subsection (2) above health and safety regulations ... may provide for any defence specified in the regulations to be available in any action for breach of that duty.

> *(3) No defence under s15(6)(b) can be pleaded in civil proceedings whether brought under s47(2) or not BUT as regards any DUTY imposed health & safety regs. may provide for any defence specified in the regs. to be available for a breach of statutory duty action.*
>
> *Thus s15(6)(b) can only be used for criminal defences BUT if duties are imposed under the regs. a defence at civil law may be included in the regs. for breach of statutory duty.*

(4) Subsections (1)(a) and (2) above are without prejudice to any right of action which exists apart from the provisions of this Act, and subsection (3) above is without prejudice to any defence which may be available apart from the provisions of the regulations there mentioned.

> *(4) s47(1)(a)–32) are without prejudice to any other right of civil action or defence to such action that parties may have by Common law or other statutes or regulations.*
> *Thus, even though a person may NOT be able to sue under this Act/regulations at civil law other provisions allowing him to sue include:*
>
> ***Common Law***
> *Negligence*
> *and/or*
> *breach*
> *of contract*
> *Breach of Occupiers Liability Acts 1957 and 1984*
> *etc.*

(5) Any term of an agreement which purports to exclude or restrict the operation of subsection (2) above, or any liability arising by virtue of that subsection, shall be void, except in so far as health and safety regulations ... provide otherwise.

> *(5) Any exclusion clause in respect of s47(2) seeking to exclude or restrict liability shall be void unless regulations provide otherwise.*

(6) In this section "damage" includes the death of, or injury to, any person (including any disease and any impairment of a person's physical or mental condition).

> *(6) In s47 'damage' includes death, injury, disease, impairment of physical or mental conditions.*

48 Application to Crown

(1) Subject to the provisions of this section, the provisions of this Part, except sections 21 to 25 and 33 to 42, and of regulations made under this Part shall bind the Crown.

> *(1) The provision of Part 1 (except sections 21 to 25 & 33 to 42) and of regulations shall bind the Crown subject to other provisions in this section.*
>
> *Thus, the Crown (inc. Government departments, QUANGOs etc.) are expected to comply with the Act/regulations but NOT the enforcement provisions or provisions relating to penalties. This is encompassed in the mediaeval maxim which still persists that, 'The Queen can do no wrong'.*

(2) Although they do not bind the Crown, sections 33 to 42 shall apply to persons in the public service of the Crown as they apply to other persons.

> *(2) sections 33 to 42 (offences) can apply to civil servants etc. even though the Crown cannot be prosecuted.*

(3) For the purposes of this Part and regulations made thereunder persons in the service of the Crown shall be treated as employees of the Crown whether or not they would be so treated apart from this subsection.

> *(3) Civil servants are deemed to be 'employees' for the purposes of this Act/regulations even though their usual status is not that of an employee but a 'Crown servant'.*

(4) Without prejudice to section 15(5), the Secretary of State may, to the extent that it appears to him requisite or expedient to do so in the interests of the safety of the State or the safe custody of persons lawfully detained, by order exempt the Crown either generally or in particular respects from all or any of the provisions of this Part which would, by virtue of subsection (1) above, bind the Crown.

(4) Without prejudice to s15(5) the SoS may exempt the Crown (in whole or in part) from any of Part 1 or regulations (despite s48(1)) if he feels such action is needed or is expedient in the interests of the safety of the State or the safe custody of persons lawfully imprisoned.

(5) The power to make orders under this section shall be exercisable by statutory instrument, and any such order may be varied or revoked by a subsequent order.

(5) SoS orders under s48 (4) may be made by statutory instrument which may be varied or revoked by a later SI.

(6) Nothing in this section shall authorise proceedings to be brought against Her Majesty in her private capacity, and this subsection shall be construed as if section 38(3) of the Crown Proceedings Act 1947 (interpretation of references in that Act to Her Majesty in her private capacity) were contained in this Act.

(6) Nothing in s48 authorises any action (civil or criminal) to be brought against the Monarch personally. Thus, the mediaeval rule that, 'The Queen can do no wrong' is preserved.

49 Adaptation of enactments to metric units or appropriate metric units

(1) [Regulations made under this subsection may amend]–

(a) any of the relevant statutory provisions; or

(b) any provision of an enactment which relates to any matter relevant to any of the general purposes of this Part but is not among the relevant statutory provisions; or

(c) any provision of an instrument made or having effect under any such enactment as is mentioned in the preceding paragraph,

by substituting an amount or quantity expressed in metric units for an amount or quantity not so expressed or by substituting an amount or quantity expressed in metric units of a description specified in the regulations for an amount or quantity expressed in metric units of a different description.

Section 49 – Adaptation of enactments to metric units or appropriate metric units

(1) Regulations made under s49(1) may amend–

(a) the Act or any regulations under the Act; or

(b) any provision of an Act of Parliament which relate any of the general purposes of Part 1 of this Act even if it is not a piece of health & safety legislation; or

(c) any provisions of a statutory instrument made under such Act as in (b), by substituting metric measurements or quantities for imperial or other measurements.

(2) The amendments shall be such as to preserve the effect of the provisions mentioned except to such extent as in the opinion of the [authority making the regulations] is necessary to obtain amounts expressed in convenient and suitable terms.

(2) Any amendments under s49(1) shall be such as to preserve the legal effect of those provisions except to such extent as SoS/MAFF shall need to express imperial and other measures in suitable and convenient metric terms (i.e. no other amendments but the measurements).

(3) Regulations made ... under this subsection may, in the case of a provision which falls within any of paragraphs (a) to (c) of subsection (1) above and contains words which refer to units other than metric units, repeal those words [if the authority making the regulations] is of the opinion that those words could be omitted without altering the effect of that provision.

(3) Regulations made under s49(1)(a) to (c) which contain words referring to units other than metric units may repeal those words if the omission of those words does not alter the overall effect of that provision, in the opinion of the SoS/MAFF making them.

[(4) The power to make regulations under this section shall be exercisable by the Secretary of State, the Minister of Agriculture, Fisheries and Food or the Secretary of State and that Minister acting jointly.]

(4) The power to make regulations under section 49(1) is exercisable by, the SoS, MAFF or both of them acting jointly.

N.B. Many regulations have been made under this section.

50   Regulations under the relevant statutory provisions

[(1) Where any power to make regulations under any of the relevant statutory provisions is exercisable by the Secretary of State, the Minister of Agriculture, Fisheries and Food or both of them acting jointly that power may be exercised either so as to give effect (with or without modifications) to proposals submitted by the Commission under section 11(2)(d) or independently of any such proposals; but the authority who is to exercise the power shall not exercise it independently of proposals from the Commission unless he has consulted the Commission and such other bodies as appear to him to be appropriate.]

Section 50 – Regulations under the relevant statutory provisions
(1) Where the SoS and/or MAFF have the power to make regulations (e.g. by s15 or 49) that power shall be exercised so as to implement proposals (with or without modifications) made by the HSC under s11(2)(d) or independently. However, SoS and/or MAFF shall not exercise independent powers unless they have consulted HSC and other appropriate bodies.

(2) Where the [authority who is to exercise any such power as is mentioned in subsection (1) above proposes to exercise that power] so as to give effect to any such proposals as are there mentioned with modifications, he shall, before making the regulations, consult the Commission.

(2) Where the SOS and/or MAFF propose to exercise any powers to implement HSC proposal with modifications they must first consult the HSC.

(3) Where the Commission proposes to submit [under section 11(2)(d)] any such proposals as are mentioned in subsection (1) above except proposals for the making of regulations under section 43(2), it shall, before so submitting them, consult–

(a) any government department or other body that appears to the Commission to be appropriate (and, in particular, in the case of proposals for the making of regulations under section 18(2), any body representing local authorities that so appears, and, in the case of proposals for the making of regulations relating to electro-magnetic radiations, the National Radiological Protection Board);

(b) such government departments and other bodies, if any, as, in relation to any matter dealt with in the proposals, the Commission is required to consult under this subsection by virtue of directions given to it by the Secretary of State.

(3) Where HSC proposes to submit under s11(2)(d) any proposals mentioned in s50(1) (except proposals for regulations under s43(2)) it shall before submitting them, consult,

(a) any government department or other appropriate body (especially LA bodies in respect of proposals under s18(2) and the National Radiological Protection Board (NRPB) in respect of proposals relating to electro magnetic radiations.)

(b) any govt. depts or other appropriate body in relation to any matter in the proposals where HSC is required to consult under s50(1)(3) by SoS directions.

(4), (5)...

51   Exclusion of application to domestic employment

Nothing in this Part shall apply in relation to a person by reason only that he employs another, or is himself employed, as a domestic servant in a private household.

Section 51 – Exclusion of application to domestic employment
Part 1 of the Act does NOT apply to domestic servants employed in private households even though they are employees.
Thus, s2 in particular does NOT apply to domestic servants (e.g. butlers, maids, etc).

[51A Application of Part to police]

[(1) For the purposes of this Part, a person who, otherwise than under a contract of employment, holds the office of constable or an appointment as police cadet shall be treated as an employee of the relevant officer.

Section 51A – Application of part to police
(1) For Part 1 purposes, police constables and cadets (who are NOT 'employee's') shall be deemed to be employees of the Chief Officer of police for the purposes of the Act.
(NB. Police were added in by the Police (Health & safety) Act 1997 but SoS has to bring in regulations to bring it into effect)

(2) In this section "the relevant officer"–

(a) in relation to a member of a police force or a special constable or police cadet appointed for a police area, means the chief officer of police,

(b) in relation to a person holding office under section 9(1)(b) or 55(1)(b) of the Police Act 1997 (police members of the National Criminal Intelligence Service and the National Crime Squad) means the Director General of the

National Criminal Intelligence Service or, as the case may be, the Director General of the National Crime Squad, and

(c) in relation to any other person holding the office of constable or an appointment as police cadet, means the person who has the direction and control of the body of constables or cadets in question.

> *(2) (a) The 'employer' of police constables (inc. special constables) and cadets is deemed to be the chief officer for a police area,*
>
> *(b) The 'employer' of police members of the National Crime Intelligence Service and the National Crime Squad is deemed to be the Director General of those bodies,*
>
> *(c) The 'employer' of any other police constable or cadet is deemed to be the person who has direction and control over such officers.*

(3) For the purposes of regulations under section 2(4) above–

(a) the Police Federation for England and Wales shall be treated as a recognised trade union recognised by each chief officer of police in England and Wales,

(b) the Police Federation for Scotland shall be treated as a recognised trade union recognised by each chief officer of police in Scotland, and

(c) any body recognised by the Secretary of State for the purposes of section 64 of the Police Act 1996 shall be treated as a recognised trade union recognised by each chief officer of police in England, Wales and Scotland.

> *(3) For the purposes of regulations made under s2(4) (trade union safety representatives)–*
>
> *(a) the Police Federation for England and Wales is deemed to be a 'recognised trade union' by each chief officer of police*
>
> *(b) the Police Federation for Scotland is deemed to be a 'recognised trade union' by each chief officer of police*
>
> *(c) any body recognised by the Home Secretary for the purposes of s64 Police Act 1996 shall be treated as a 'recognised trade union' recognised by a specified person.*

(4) Regulations under section 2(4) above may provide, in relation to persons falling within subsection (2)(b) or (c) above, that a body specified in the regulations is to be treated as a recognised trade union recognised by such person as may be specified.]

> *(4) Regulations under s2(4) may provide in relation to persons mentioned in s51(2)(b) or (c) that a body treated as a 'recognised trade union' shall be recognised by such person as may be specified.*

52 Meaning of work and at work

(1) For the purposes of this Part–

    (a) "work" means work as an employee or as a self-employed person;

    (b) an employee is at work throughout the time when he is in the course of his employment, but not otherwise;

    [(bb) a person holding the office of constable is at work throughout the time when he is on duty, but not otherwise; and]

    (c) a self-employed person is at work throughout such time as he devotes to work as a self-employed person;

and, subject to the following subsection, the expressions "work" and "at work", in whatever context, shall be construed accordingly.

Section 52 – Meaning of work and at work

(1) Wherever it appears in Part 1 (section 1 to 54)–

(a) 'work' means work as an employee or self employed person;

(b) an employee is at work when he is doing what he is paid to do during his working hours. Personal acts and 'frolics of his own' in employers time are NOT included & employers may not be liable for these acts.

(bb) police constables are at work at all times on duty but not otherwise; and

(c) a self-employed person is at work throughout such time as he devotes to work as a self-employed person. This probably includes time spent in the evenings doing the accounts etc.

'Work' and 'at work' are to be interpreted accordingly subject to s52(2).

(2) Regulations made under this subsection may–

    (a) extend the meaning of "work" and "at work" for the purposes of this Part; and

    (b) in that connection provide for any of the relevant statutory provisions to have effect subject to such adaptations as may be specified in the regulations.

(2) Regulations made under s52 may–

(a) extend the meaning of 'work' and 'at work' for Part 1 purposes and

(b) provide for the Act/regulations to have effect subject to adaptations in the regulations.

[(3) The power to make regulations under subsection (2) above shall be exercisable by the Secretary of State, the Minister of Agriculture, Fisheries and Food or the Secretary of State and that Minister acting jointly.]

(3) The power to make regulations under s52(2) belongs to the SoS and/or MAFF.

53 General interpretation of Part I

(1) In this Part, unless the context otherwise requires–

...

"article for use at work" means–

(a) any plant designed for use or operation (whether exclusively or not) by persons at work, and

(b) any article designed for use as a component in any such plant;

["article of fairground equipment" means any fairground equipment or any article designed for use as a component in any such equipment.]

"code of practice" (without prejudice to section 16(8)) includes a standard, a specification and any other documentary form of practical guidance;

"the Commission" has the meaning assigned by section 10(2);

"conditional sale agreement" means an agreement for the sale of goods under which the purchase price or part of it is payable by instalments, and the property in the goods is to remain in the seller (notwithstanding that the buyer is to be in possession of the goods) until such conditions as to the payment of instalments or otherwise as may be specified in the agreement are fulfilled;

"contract of employment" means a contract of employment or apprenticeship (whether express or implied and, if express, whether oral or in writing);

"credit-sale agreement" means an agreement for the sale of goods, under which the purchase price or part of it is payable by instalments, but which is not a conditional sale agreement;

["customs officer" means an officer within the meaning of the Customs and Excise Management Act 1979;]

"domestic premises" means premises occupied as a private dwelling (including any garden, yard, garage, outhouse or other appurtenance of such premises which is not used in common by the occupants of more than one such dwelling), and "non-domestic premises" shall be construed accordingly;

"employee" means an individual who works under a contract of employment [or is treated by section 51A as being an employee], and related expressions shall be construed accordingly;

"enforcing authority" has the meaning assigned by section 18(7);

"the Executive" has the meaning assigned by section 10(5);

"the existing statutory provisions" means the following provisions while and to the extent that they remain in force, namely the provisions of the Acts mentioned in Schedule 1 which are specified in the third column of that Schedule and of the regulations, orders or other instruments of a legislative character made or having effect under any provision so specified;

...

["fairground equipment" means any fairground ride, any similar plant which is designed to be in motion for entertainment purposes with members of the public on or inside it or any plant which is designed to be used by members of the public for entertainment purposes either as a slide or for bouncing upon, and in this definition the reference to plant which is designed to be in motion with members of the public on or inside it includes a reference to swings, dodgems and other plant which is designed to be in motion wholly or partly under the control of, or to be put in motion by, a member of the public;]

"the general purposes of this Part" has the meaning assigned by section 1;

"health and safety regulations" has the meaning assigned by section 15(1);

"hire-purchase agreement" means an agreement other than a conditional sale agreement, under which–

(a) goods are bailed or (in Scotland) hired in return for periodical payments by the person to whom they are bailed or hired; and

(b) the property in the goods will pass to that person if the terms of the agreement are complied with and one or more of the following occurs:

    (i)   the exercise of an option to purchase by that person;

    (ii) the doing of any other specified act by any party to the agreement;

    (iii) the happening of any other event;

and "hire-purchase" shall be construed accordingly;

"improvement notice" means a notice under section 21;

"inspector" means an inspector appointed under section 19;

...

"local authority" means–

(a) in relation to England ... , a county council, ..., a district council, a London borough council, the Common Council of the City of London, the Sub-Treasurer of the Inner Temple or the Under-Treasurer of the Middle Temple,

[(aa) in relation to Wales, a county council or a county borough council,]

(b) in relation to Scotland, a [council constituted under section 2 of the Local Government etc (Scotland) Act 1994];

["micro-organism" includes any microscopic biological entity which is capable of replication;]

"offshore installation" means any installation which is intended for underwater exploitation of mineral resources or exploration with a view to such exploitation;

"personal injury" includes any disease and any impairment of a person's physical or mental condition;

"plant" includes any machinery, equipment or appliance;

"premises" includes any place and, in particular, includes–

(a) any vehicle, vessel, aircraft or hovercraft,

(b) any installation on land (including the foreshore and other land intermittently covered by water), any offshore installation, and any other installation (whether floating, or resting on the seabed or the subsoil thereof, or resting on other land covered with water or the subsoil thereof), and

(c) any tent or movable structure;

"prescribed" means prescribed by regulations made by the Secretary of State;

"prohibition notice" means a notice under section 22;

...

"the relevant statutory provisions" means–

(a) the provisions of this Part and of any health and safety regulations ...; and

(b) the existing statutory provisions;

"self-employed person" means an individual who works for gain or reward otherwise than under a contract of employment, whether or not he himself employs others;

"substance" means any natural or artificial substance [(including micro-organisms)], whether in solid or liquid form or in the form of a gas or vapour;

...

"supply", where the reference is to supplying articles or substances, means supplying them by way of sale, lease, hire or hire-purchase, whether as principal or agent for another.

(2)(6) ...

> **Section 53 – General Interpretation of Part 1**
> *This section is a mandatory legal guide to phrases used in Part 1 (ss1 to 54) of the Act and is fairly self explanatory.*
> *Some explanation of some terms has been given above under appropriate sections.*
> *Further explanations are given in the other legislation referred to or case law where interpretation disputes have arisen.*

54 Application of Part I to Isles of Scilly

This Part, in its application to the Isles of Scilly, shall apply as if those Isles were a local government area and the Council of those Isles were a local authority.

> **Section 54 – Application of Part 1 to Isles of Scilly**
> *Part 1 applies to the Isles of Scilly as if it the Council of the Isles of Scilly a local government unit.*

PART II

The Employment Medical Advisory Service

55 Functions of, and responsibility for maintaining, employment medical advisory service

(1) There shall continue to be an employment medical advisory service, which shall be maintained for the following purposes, that is to say–

  (a) securing that the Secretary of State, the Health and Safety Commission ... and others concerned with the health of employed persons or of persons seeking or training for employment can be kept informed of, and adequately advised on, matters of which they ought respectively to take cognisance concerning the safeguarding and improvement of the health of those persons;

  (b) giving to employed persons and persons seeking or training for employment information and advice on health in relation to employment and training for employment;

  (c) other purposes of the Secretary of State's functions relating to employment.

> **PART II – THE EMPLOYMENT MEDICAL ADVISORY SERVICE (EMAS)**
> **Section 55 – Functions of and responsibility for maintaining EMAS**
> *(1) EMAS continues (see EMAS Act, 1972) and has the following purposes:*
> *(a) keeping the following informed & adequately advised on matters which they ought to take account of in safeguarding & improving the health of such person:*
> *SOS, HSC & others concerned with the health & safety of employed persons and those seeking or training for employment.*
> *(b) giving to employees and trainees (inc. those seeking training) advice on health in relation to employment and training. Any employee/trainee can seek this advice.*
> *(c) other purposes of SOS functions relating to employment.*

(2) The authority responsible for maintaining the said service shall be the Secretary of State; but if arrangements are made by the Secretary of State for that responsibility to be discharged on his behalf by the Health and Safety Commission or some other body, then, while those arrangements operate, the body so discharging that responsibility (and not the Secretary of State) shall be the authority responsible for maintaining that service.

> *(2) Overall responsibility for EMAS is with SOS but SOS may delegate this function to HSC or some other body in which case that body has responsibilities for the time being.*

(3) The authority for the time being responsible for maintaining the said service may also for the purposes mentioned in subsection (1) above, and for the purpose of assisting employment medical advisers in the performance of their functions, investigate or assist in, arrange for or make payments in respect of the investigation of problems arising in connection with any such matters as are so mentioned or otherwise in connection with the functions of employment medical advisers, and for the purpose of investigating or assisting in the investigation of such problems may provide and maintain such laboratories and other services as appear to the authority to be requisite.

> *(3) The body responsible for EMAS for the time being may exercised s55(1) powers & for the purpose of assisting EMAS advisors may:*
> - *investigate or assist in investigating*
> - *arrange or make payments for investigation, functions of EMAS advisers, etc.*
> - *provide such laboratories & other services as appear to be needed*

(4) Any arrangements made by the Secretary of State in pursuance of subsection (2) above may be terminated by him at any time, but without prejudice to the making of other arrangements at any time in pursuance of that subsection (including arrangements which are to operate from the time when any previous arrangements so made cease to operate).

> *(4) SOS may terminate arrangements at any time under s55(2) without prejudice to the making of other arrangements immediately.*

(5) Without prejudice to sections 11(4)(a) and 12(b), it shall be the duty of the Health and Safety Commission, if so directed by the Secretary of State, to enter into arrangements with him for the Commission to be responsible for maintaining the said service.

> *(5) HSC has a duty to enter into arrangements with the SOS for HSC to be responsible for EMAS, without prejudice to s11(4)(a) (delegation to HSE) & 12(b) (SOS directions to HSC).*

(6) In subsection (1) above–

    (a) the reference to persons training for employment shall include persons attending industrial rehabilitation courses provided by virtue of the Employment and Training Act 1973; and

    (b) the reference to persons (other than the Secretary of State and the [Health and Safety Commission]) concerned with the health of employed persons or of persons seeking or training for employment shall be taken to include organisations representing employers, employees and occupational health practitioners respectively.

(6) In s55(1)

(a) trainees include person attending industrial rehabilitation courses by virtue of the Employment & Training Act, 1973; and

(b) reference to persons other than SOS, HSC shall include:

- *employers organisations (e.g. CBI & Chambers of Commerce)*
- *employees organisations (e.g. TUC and trade unions and staff associations)*
- *occupational health practitioners.*

56 Functions of authority responsible for maintaining the service

(1) The authority for the time being responsible for maintaining the employment medical advisory service shall for the purpose of discharging that responsibility appoint persons to be employment medical advisers, and may for that purpose appoint such other officers and servants as it may determine, subject however to the requisite approval as to numbers, that is to say–

    (a) where that authority is the Secretary of State, the approval of the Minister for the Civil Service;

    (b) otherwise, the approval of the Secretary of State given with the consent of that Minister.

Section 56 – Functions of authority responsible for maintaining EMAS

(1) The authority for the time being responsible for EMAS in order to discharge that function shall:

- *appoint persons to be EMAS advisers (i.e. medical doctors)*
- *appoint such other officers/servants they may determine*

subject to the approval of numbers by Minister of Civil Service (now Treasury) if the authority is the SOS or the SOS if the authority is some other body (e.g. HSC, HSE etc.).

(2) A person shall not be qualified to be appointed, or to be, an employment medical adviser unless he is a fully registered medical practitioner.

> *(2) EMAS employment medical advisors must be fully registered medical practitioners.*

(3) The authority for the time being responsible for maintaining the said service may determine the cases and circumstances in which the employment medical advisers or any of them are to perform the duties or exercise the powers conferred on employment medical advisers by or under this Act or otherwise.

> *(3) The authority for the time being for EMAS may decide the cases and circumstances in which employment medical advisers are to perform duties or exercise powers by statute.*

(4) Where as a result of arrangements made in pursuance of section 55(2) the authority responsible for maintaining the said service changes, the change shall not invalidate any appointment previously made under subsection (1) above, and any such appointment subsisting when the change occurs shall thereafter have effect as if made by the new authority.

> *(4) Where under s55(2) the authority for EMAS changes, the change shall not invalidate pervious appointments & shall operate as if made by the new authority.*

57 Fees

(1) The Secretary of State may by regulations provide for such fees as may be fixed by or determined under the regulations to be payable for or in connection with the performance by the authority responsible for maintaining the employment medical advisory service of any function conferred for the purposes of that service on that authority by virtue of this Part or otherwise.

> **Section 57 – Fees**
> *(1) The SOS may make regulations laying down fees for functions performed by the authority responsible for EMAS. People can be charged for EM'S services.*

(2) For the purposes of this section, the performance by an employment medical adviser of his functions shall be treated as the performance by the authority responsible for maintaining the said service of functions conferred on that authority as mentioned in the preceding subsection.

> *(2) The performance of functions by EMAS advisors is treated as performance by the responsible authority for fees purposes.*

(3) The provisions of subsections (4), (5) and (8) of section 43 shall apply in relation to regulations under this section with the modification that references to subsection (2) of that section shall be read as references to subsection (1) of this section.

> *(3) S43(4)(5)&(8) (Financial provisions) shall apply to regulations made under s57 with the modification that reference to s43(2) shall be read as references to s57(1).*

(4) Where an authority other than the Secretary of State is responsible for maintaining the said service, the Secretary of State shall consult that authority before making any regulations under this section.

> *(4) Where bodies other than the SOS are responsible for EMAS (e.g. HSC, HSE) the SOS shall consult those bodies before making regulations under s57.*

58 Other financial provisions

(1) The authority for the time being responsible for maintaining the employment medical advisory service may pay-

    (a) to employment medical advisers such salaries or such fees and travelling or other allowances; and

    (b) to other persons called upon to give advice in connection with the execution of the authority's functions under this Part such travelling or other allowances or compensation for loss of remunerative time; and

    (c) to persons attending for medical examinations conducted by, or in accordance with arrangements made by, employment medical advisers (including pathological, physiological and radiological tests and similar investigations so conducted) such travelling or subsistence allowances or such compensation for loss of earnings,

as the authority may, with the requisite approval, determine.

> ### Section 58 – Other financial provisions
>
> *(1) The EMAS authority may (subject to requisite approval) pay:*
>
> *(a) to employment medical advisers salaries, fees & other expenses; and*
>
> *(b) to other advisers fees, expenses and compensation for loss of paid time; and*
>
> *(c) to persons attending for medical examinations (inc. pathological, physiological, radiological, etc) travelling & subsistence expenses & compensation for loss of earnings.*

(2) For the purposes of the preceding subsection the requisite approval is–

    (a) where the said authority is the Secretary of State, the approval of the Minister for the Civil Service;

    (b) otherwise, the approval of the Secretary of State given with the consent of that Minister.

(2) Powers under s58(1) must be exercised with the approval of:

(a) Minister of Civil Service (Treasury) where SOS is the EMAS authority

(b) SOS where other bodies are the EMAS authority.

(3) Where an authority other than the Secretary of State is responsible for maintaining the said service, it shall be the duty of the Secretary of State to pay to that authority such sums as are approved by the Treasury and as he considers appropriate for the purpose of enabling the authority to discharge that responsibility.

(3) Where other bodies (NOT the SOS) are the EMAS authority it is SOS's duty to pay to such bodies sums approved by the Treasury as he considers appropriate for the discharge of that responsibility.

59 Duty of responsible authority to keep accounts and to report

(1) It shall be the duty of the authority for the time being responsible for maintaining the employment medical advisory service–

    (a) to keep, in relation to the maintenance of that service, proper accounts and proper records in relation to the accounts;

    (b) to prepare in respect of each accounting year a statement of accounts relating to the maintenance of that service in such form as the Secretary of State may direct with the approval of the Treasury; and

    (c) to send copies of the statement to the Secretary of State and the Comptroller and Auditor General before the end of the month of November next following the accounting year to which the statement relates.

Section 59 – Duty of responsible person to keep accounts and to report

(1) the EMAS authority for the time being shall:

(a) keep proper accounts and records relating to EMAS;

(b) prepare an EMAS annual statement each accounting year; and

(c) send copies of the statement to the SOS and Comptroller & Auditor General (CAAG) by the end of Nov. next following the accounting year to which the statement relates.

(2)   The Comptroller and Auditor General shall examine, certify and report on each statement received by him in pursuance of subsection (1) above and shall lay copies of each statement and of his report before each House of Parliament.

(2) CAAG shall examine, certify & report on each statement received under s59(1) and shall lay copies before both House of Commons and House of Lords.

(3)   It shall also be the duty of the authority responsible for maintaining the employment medical advisory service to make to the Secretary of State, as soon as possible after the end of each accounting year, a report on the discharge of its responsibilities in relation to that service during that year; and the Secretary of State shall lay before each House of Parliament a copy of each report made to him in pursuance of this subsection.

(3) the EMAS authority must report to SOS as soon as possible after the end of each accounting year on the discharge of EMAS responsibilities & that report shall be laid before the House of Commons and House of Lords.

(4)   Where as a result of arrangements made in pursuance of section 55(2) the authority responsible for maintaining the employment medical advisory service changes, the change shall not affect any duty imposed by this section on the body which was responsible for maintaining that service before the change.

(4) Where changes in EMAS authority occur by s55(2), the change shall not affect the duties in s59 & previous authorities must still comply.

(5)   No duty imposed on the authority for the time being responsible for maintaining the employment medical advisory service by subsection (1) or (3) above shall fall on the Commission (which is subject to corresponding duties under Schedule 2) or on the Secretary of State.

(5) No duty under s59(1) or (3) shall fall on HSC (which has similar duties in Schedule 2) or on the SOS. Thus, it seems that HSE will have the main duties.

(6)   In this section "accounting year" means, except so far as the Secretary of State otherwise directs, the period of twelve months ending with 31st March in any year.

(6) In s59 'accounting year' means the period ending 31st March in any year, unless SOS directs otherwise.

60 Supplementary

(1) It shall be the duty of the Secretary of State to secure that each [Health Authority arranges for] a fully registered medical practitioner to furnish, on the application of an employment medical adviser, such particulars of the school medical record of a person who has not attained the age of eighteen and such other information relating to his medical history as the adviser may reasonably require for the efficient performance of his functions; but no particulars or information about any person which may be furnished to an adviser in pursuance of this subsection shall (without the consent of that person) be disclosed by the adviser otherwise than for the efficient performance of his functions.

> **Section 60 – Supplementary**
> *(1) It is the SOS's duty to secure that each (Local) Health Authority (LHA) arranges for a fully registered medical practitioner to furnish, on the application of EMAS employment medical advisers, particulars of the school medical record of a person who has not reached 18 & such other reasonable information to enable the efficient discharge of his functions. The EMAS employment medical adviser may NOT disclose any such information to others (except for the purposes of his functions) to anyone else without the consent of the patient.*

(2) In its application to Scotland the preceding subsection shall have effect with the substitution of the words "every Health Board arrange for one of their" for the words from "each" to "its".

> *(2) Substitutes 'every Health Board' for 'every health authority' in s80(1) as applied to Scotland.*

(3) ...

> *(3) This subsection was repealed by Statute Law (Repeals) Act, 1986.*

(4) References to the chief employment medical adviser or a deputy chief employment medical adviser in any provision of an enactment or instrument made under an enactment shall be read as references to a person appointed for the purposes of that provision by the authority responsible for maintaining the employment medical advisory service.

> *(4) References to 'chief employment medical adviser' (CEMA) or 'deputy chief employment medical adviser' (DCMEA) under any Act/regulation shall be deemed to refer to a person appointed by an EMAS authority for the time being.*

(5) The following provisions of the Employment Medical Advisory Service Act 1972 (which are superseded by the preceding provisions of this Part or rendered unnecessary by provisions contained in Part I), namely sections 1 and 6 and Schedule 1, shall cease to have effect; but-

(a) in so far as anything done under or by virtue of the said section 1 or Schedule 1 could have been done under or by virtue of a corresponding provision of Part I or this Part, it shall not be invalidated by the repeal of that section and Schedule by this Act but shall have effect as if done under or by virtue of that corresponding provision; and

(b) any order made under the said section 6 which is in force immediately before the repeal of that section by this Act shall remain in force notwithstanding that repeal, but may be revoked or varied by regulations under section 43(2) or 57, as if it were an instrument containing regulations made under section 43(2) or 57 as the case may require.

(5) Sections 1 to 6 and Schedule 1 of Employment Medical Advisory Service Act, 1972 cease to have effect as they are replaced by ss55 to 60 and Part 1 of this Act.
However,
(a) past acts are still valid & are deemed to have been exercised under this Act.
(b) past regulations made under the 1972 Act remain in force under this Act but may be revoked or varied by regulation making powers under s43(2) or 57.

(6) Where any Act (whether passed before, or in the same Session as, this Act) or any document refers, either expressly or by implication, to or to any enactment contained in any of the provisions of the said Act of 1972 which are mentioned in the preceding subsection, the reference shall, except where the context otherwise requires, be construed as, or as including, a reference to the corresponding provision of this Act.

(6) Where any Act or document (preceding this Act) refers specifically to provisions of the 1972 Act (now repealed) they shall be deemed to be references to corresponding provisions (in ss55 to 60) of this Act, unless the context requires otherwise.

(7) Nothing in subsection (5) or (6) above shall be taken as prejudicing the operation of section 38 of the Interpretation Act 1889 (which relates to the effect of repeals).

(7) Nothing in s60(5) or (6) shall prejudice s38 Interpretation Act,1889 (now Interpretation Act,1978 ss16(1) & 17(2)) (relating to the effect of repeals).

NB. EMAS may be contacted via HSE offices and may advise employers and employees and others on any medical problem to do with work including helping to track chemical problems, work related musculoskeletal disorders, stress, etc etc.

A pamphlet is available on EMAS services from HSE.

PART III BUILDING REGULATIONS AND AMENDMENT OF BUILDING (Scotland) Act 1959.

As this Part (SECTIONS 61 TO 76) is applicable to Building Control and is not of general application no explanatory notes are given of these sections. Reference should be made to appropriate Building legislation.

PART IV

MISCELLANEOUS AND GENERAL

Section 77 – Amendment of Radiological Protection Act,1970

This section merely included a new section 6a and 7a in the radiological protection act, 1970 and other amendments; its now repealed.

78   Amendment of Fire Precautions Act 1971

(1)–(9) …

(10) Schedule 8 (transitional provisions with respect to fire certificates under the Factories Act 1961 or the Offices, Shops and Railway Premises Act 1963) shall have effect.

Section 78 – Amendment of Fire Precautions Act, 1971.

This section merely amends and adds sections to the Fire Precautions Act, 1971
NB. The Workplace (Fire Precautions) Regulations, 1997 now impose new requirements alongside the FPA.

79 – Amendment of Companies Act as to directors reports.

Now "spent"

Section 79 – Amendment of Companies Act as to directors reports.

This section merely amended Companies legislation to include reference in directors reports of health and safety matters. Companies legislation now contains this provision.

80 General power to repeal or modify Acts and instruments

(1) Regulations made under this subsection may repeal or modify any provision to which this subsection applies if it appears to the authority making the regulations that the repeal or, as the case may be, the modification of that provision is expedient in consequence of or in connection with any provision made by or under Part I.

> **Section 80 – General Power to Repeal or modify Acts and instruments.**
> *(1) Regulations made under s80(1) may repeal or modify any provision to which s80(1) applies if SoS and/or MAFF think it expedient relating to any matter under Part 1.*

(2) Subsection (1) above applies to any provision, not being among the relevant statutory provisions, which–

    (a) is contained in this Act or in any other Act passed before or in the same Session as this Act; or

    (b) is contained in any regulations, order or other instrument of a legislative character which was made under an Act before the passing of this Act; or

    (c) applies, excludes or for any other purpose refers to any of the relevant statutory provisions and is contained in any Act not falling within paragraph (a) above or in any regulations, order or other instrument of a legislative character which is made under an Act but does not fall within paragraph (b) above.

> *(2) s80(1) applies to any provision (not in the 'relevant statutory provisions'–see s53(1) which–*
>
> *(a) is contained in this Act Parts II to IV or any other Act passed before or in the same session as this Act; or*
>
> *(b) contained in regulations under Acts made before this Act*
>
> *(c) other legislation referring to the 'relevant statutory provisions' not covered by (b)*

[(2A) Subsection (1) above shall apply to provisions in [the Employment Rights Act 1996 or the Trade Union and Labour Relations (Consolidation) Act 1992 which derive from provisions of the Employment Protection (Consolidation) Act 1978 which re-enacted] provisions previously contained in the Redundancy Payments Act 1965, the Contracts of Employment Act 1972 and the Trade Union and Labour Relations Act 1974 as it applies to provisions contained in Acts passed before or in the same Session as this Act.]

(2A) s80(1) shall apply to provisions in the Employment Rights Act 1996 to Trade Union and Labour Relations (Consolidation) Act 1992 provided they consolidated matters from the earlier legislation mentioned.

(3) Without prejudice to the generality of subsection (1) above, the modifications which may be made by regulations thereunder include modifications relating to the enforcement of provisions to which this section applies (including the appointment of persons for the purpose of such enforcement, and the powers of persons so appointed).

(3) Modifications made under s80(1) regulations may include references to enforcement & appointment of inspectors and their powers without prejudice to other matters covered by s80(1).

[(4) The power to make regulations under subsection (1) above shall be exercisable by the Secretary of State, the Minister of Agriculture, Fisheries and Food or the Secretary of State and that Minister acting jointly; but the authority who is to exercise the power shall, before exercising it, consult such bodies as appear to him to be appropriate.

(4) The power to make s80(1) regulations is exercisable by the SoS and/or MAFF subject to appropriate consultation beforehand.

(5) In this section "the relevant statutory provisions" has the same meaning as in Part I.]

(5) 'relevant statutory provisions' has the same meaning as in s53(1).

81 Expenses and receipts

There shall be paid out of money provided by Parliament-

(a) any expenses incurred by a Minister of the Crown or government department for the purposes of this Act; and

(b) any increase attributable to the provisions of this Act in the sums payable under any other Act out of money so provided;

and any sums received by a Minister of the Crown or government department by virtue of this Act shall be paid into the Consolidated Fund.

Section 81 - Expenses and receipts
Out of money provided by Parliament there shall be paid—
(a) Any expenses of Govt Ministers or Depts for the purposes of this Act; and
(b) any increase that this Act is responsible for under any other Act.
Any sums received shall be paid into the Consolidated Fund.

82 General provisions as to interpretation and regulations

(1) In this Act-

    (a) "Act" includes a provisional order confirmed by an Act;

    (b) "contravention" includes failure to comply, and "contravene" has a corresponding meaning;

    (c) "modifications" includes additions, omissions and amendments, and related expressions shall be construed accordingly;

    (d) any reference to a Part, section or Schedule not otherwise identified is a reference to that Part or section of, or Schedule to, this Act.

(2) Except in so far as the context otherwise requires, any reference in this Act to an enactment is a reference to it as amended, and includes a reference to it as applied, by or under any other enactment, including this Act.

(3) Any power conferred by Part I or II or this Part to make regulations-

    (a) includes power to make different provision by the regulations for different circumstances or cases and to include in the regulations such incidental, supplemental and transitional provisions as the authority making the regulations considers appropriate in connection with the regulations; and

    (b) shall be exercisable by statutory instrument, which shall be subject to annulment in pursuance of a resolution of either House of Parliament.

Section 82 – General Provisions as to interpretation and regulations.
This is an additional interpretation section most of which is fairly self explanatory

Section 83 – Minor and consequential amendments and repeals
(1) Schedule 9 contained some minor amendments as a result of this Act to other legislation
(2) Schedule 10 contained some repeals (whole or part) to other Acts as a result of this Act.
These provisions are now repealed.

84 Extent, and application of Act

(1) This Act, except–

(a) Part I and this Part so far as may be necessary to enable regulations under section 15 ... to be made and operate for the purpose mentioned in paragraph 2 of Schedule 3; and

(b) paragraph ... 3 of Schedule 9,

does not extend to Northern Ireland.

> **Section 84 – Extent and application of Act.**
> *(1) This Act does NOT apply to Northern Ireland except*
> *(a) Part 1 & IV as are necessary to make regulations under s15 for the purposes of Schedule 3 paragraph 2 (imports into UK – see post); and paragraph 3 of Schedule 9 (Parliamentary Commissioner Act – see post).*
> *NB. The Act applies to England, Wales and Scotland but NOT to NI which has its own regulations which approximate to this Act.*

(2) Part III, except section 75 and Schedule 7, does not extend to Scotland.

> *(2) Part III of the Act (except ss75 and Schedule 7) do NOT apply to Scotland (Building regulations)*

(3) Her Majesty may by Order in Council provide that the provisions of Parts I and II and this Part shall, to such extent and for such purposes as may be specified in the Order, apply (with or without modification) to or in relation to persons, premises, work, articles, substances and other matters (of whatever kind) outside Great Britain as those provisions apply within Great Britain or within a part of Great Britain so specified.

For the purposes of this subsection "premises", "work" and "substance" have the same meaning as they have for the purposes of Part I.

> *(3) The Queen may by Order in Council (i.e. Privy Council) provide that Parts 1 and II of this Act to such extent as may be laid down may apply outside Great Britain or within a part of Great Britain.*
> *NB. This section was used to extend the Act to offshore oil and gas rigs.*

(4) An Order in Council under subsection (3) above–

(a) may make different provision for different circumstances or cases;

(b) may (notwithstanding that this may affect individuals or bodies corporate outside the United Kingdom) provide for any of the provisions mentioned in that subsection, as applied by such an Order, to apply to individuals whether or not they are British subjects and to bodies corporate whether or not they are incorporated under the law of any part of the United Kingdom;

(c) may make provision for conferring jurisdiction on any court or class of courts specified in the Order with respect to offences under Part I committed outside Great Britain or with respect to causes of action arising by virtue of section 47 (2) in respect of acts or omissions taking place outside Great Britain, and for the determination, in accordance with the law in force in such part of Great Britain as may be specified in the Order, of questions arising out of such acts or omissions;

(d) may exclude from the operation of section 3 of the Territorial Waters Jurisdiction Act 1878 (consents required for prosecutions) proceedings for offences under any provision of Part I committed outside Great Britain;

(e) may be varied or revoked by a subsequent Order in Council under this section;

and any such Order shall be subject to annulment in pursuance of a resolution of either House of Parliament.

(4) Such Order in Council may be passed flexibly as indicated in section 84(4)(a) to (e) which are fairly self explanatory.

(5) ...

(6) Any jurisdiction conferred on any court under this section shall be without prejudice to any jurisdiction exercisable apart from this section by that or any other court.

85 Short title and commencement

(1) This Act may be cited as the Health and Safety at Work etc Act 1974.

Section 85 – Short Title and Commencement
(1) The Act may be quoted as the Health & safety at Work etc. Act, 1974

(2) This Act shall come into operation on such day as the Secretary of State may by order made by statutory instrument appoint, and different days may be appointed under this subsection for different purposes.

(2) SoS had power to bring into operation various parts of the Act by statutory instrument. Most commencement orders have now been made.

(3)   An order under this section may contain such transitional provisions and savings as appear to the Secretary of State to be necessary or expedient in connection with the provisions thereby brought into force, including such adaptations of those provisions or any provision of this Act then in force as appear to him to be necessary or expedient in consequence of the partial operation of this Act (whether before or after the day appointed by the order).

(3)   transitional provisions were allowed for by order by SoS.

SCHEDULE 1

EXISTING ENACTMENTS WHICH ARE RELEVANT STATUORY PROVISIONS

Schedule 1 – *This specifies pre-1974 Statutes which are included in the term 'relevant statutory provisions' (see s53(1)).*

Chapter	Short title	Provisions which are relevant statutory provisions
1875 c 17	The Explosives Act 1875	The whole Act except sections 30 to 32, 80 and 116 to 121.
1882 c 22	The Boiler Explosions Act 1882	The whole Act.
1890 c 35	The Boiler Explosions Act 1882	The whole Act.
1906 c 14	The Alkali, &c Works Regulation Act 1906	The whole Act.
1909 c 43	The Revenue Act 1909	Section 11.
1919 c 23	The Anthrax Prevention Act 1919	The whole Act.
1920 c 65	The Employment of Women, Young Persons and Children Act 1920	The whole Act.
1922 c 35	The Celluloid and Cinematograph Film Act 1922	The whole Act.
1923 c 17	The Explosives Act 1923	The whole Act.
1926 c 43	The Public Health (Smoke Abatement) Act 1926	The whole Act.
1928 c 32	The Petroleum (Consolidation) Act 1928	The whole Act.
1936 c 22	The Hours of Employment (Conventions) Act 1936	The whole Act except section 5.
1936 c 27	The Petroleum (Transfer of Licences) Act 1936	The whole Act.

Chapter	Short title	Provisions which are relevant statutory provisions
1937 c 45	The Hydrogen Cyanide (Fumigation) Act 1937	The whole Act.
1945 c 19	The Ministry of Fuel and Power Act 1945	Section 1(1) so far as it relates to maintaining and improving the safety, health and welfare of persons employed in or about mines and quarries in Great Britain.
1946 c 59	The Coal Industry Nationalisation Act 1946	Section 42(1) and (2).
1948 c 37	The Radioactive Substances Act 1948	Section 5(1)(a).
1951 c 21	The Alkali, &c Works Regulation (Scotland) Act 1951	The whole Act.
1951 c 58	The Fireworks Act 1951	Sections 4 and 7.
1952 c 60	The Agriculture (Poisonous Substances) Act 1952	The whole Act.
1953 c 47	The Emergency Laws (Miscellaneous Provisions) Act 1953	Section 3.
[...	...	...]
1954 c 70	The Mines and Quarries Act 1954	The whole Act except section 151.
1956 c 49	The Agriculture (Safety, Health and Welfare Provisions) Act 1956	The whole Act.
1961 c 34	The Factories Act 1961	The whole Act except section 135.
1961 c 64	The Public Health Act 1961	Section 73.
1962 c 58	The Pipe-lines Act 1962	Sections 20 to 26, 33, 34 and 42, Schedule 5.

Chapter	Short title	Provisions which are relevant statutory provisions
1963 c 41	The Offices, Shops and Railway Premises Act 1963	The whole Act.
1965 c 57	The Nuclear Installations Act 1965	Sections 1, 3 to 6, 22 and [24A], Schedule 2.
1969 c 10	The Mines and Quarries (Tips) Act 1969	Sections 1 to 10.
1971 c 20	The Mines Management Act 1971	The whole Act.
1972 c 28	The Employment Medical Advisory Service Act 1972	The whole Act except sections 1 and 6 and Schedule 1.

SCHEDULE 2

ADDITIONAL PROVISIONS RELATING TO CONSTITUTION ETC, OF THE
COMMISSION AND EXECUTIVE

Schedule 2 – *Contains additional provisions relating to the Constitution of the HSC and HSE.*

Tenure of office

1. Subject to paragraphs 2 to 4 below, a person shall hold and vacate office as a member or as chairman or deputy chairman in accordance with the terms of the instrument appointing him to that office.

2. A person may at any time resign his office as a member or as chairman or deputy chairman by giving the Secretary of State a notice in writing signed by that person and stating that he resigns that office.

3.–(1) If a member becomes or ceases to be the chairman or deputy chairman, the Secretary of State may vary the terms of the instrument appointing him to be a member so as to alter the date on which he is to vacate office as a member.

(2) If the chairman or deputy chairman ceases to be a member he shall cease to be chairman or deputy chairman, as the case may be.

4.–(1) If the Secretary of State is satisfied that a member–

(a) has been absent from meetings of the Commission for a period longer than six consecutive months without the permission of the Commission; or

(b) has become bankrupt or made an arrangement with his creditors; or

(c) is incapacitated by physical or mental illness; or

(d) is otherwise unable or unfit to discharge the functions of a member,

the Secretary of State may declare his office as a member to be vacant and shall notify the declaration in such manner as the Secretary of State thinks fit; and thereupon the office shall become vacant.

(2) In the application of the preceding sub-paragraph to Scotland for the references in paragraph (b) to a member's having become bankrupt and to a member's having made an arrangement with his creditors there shall be substituted respectively references to sequestration of a member's estate having been awarded and to a member's having made a trust deed for behoof of his creditors or a composition contract.

Remuneration etc of members

5. The Commission may pay to each member such remuneration and allowances as the Secretary of State may determine.

6. The Commission may pay or make provision for paying, to or in respect of any member, such sums by way of pension, superannuation allowances and gratuities as the Secretary of State may determine.

7. Where a person ceases to be a member otherwise than on the expiry of his term of office and it appears to the Secretary of State that there are special circumstances which make it right for him to receive compensation, the Commission may make to him a payment of such amount as the Secretary of State may determine.

Proceedings

8. The quorum of the Commission and the arrangements relating to meetings of the Commission shall be such as the Commission may determine.

9. The validity of any proceedings of the Commission shall not be affected by any vacancy among the members or by any defect in the appointment of a member.

Staff

10. It shall be the duty of the Executive to provide for the Commission such officers and servants as are requisite for the proper discharge of the Commission's functions; and any reference in this Act to an officer or servant of the Commission is a reference to an officer or servant provided for the Commission in pursuance of this paragraph.

11. The Executive may appoint such officers and servants as it may determine with the consent of the Secretary of State as to numbers and terms and conditions of service.

12. The Commission shall pay to the Minister for the Civil Service, at such times in each accounting year as may be determined by that Minister subject to any directions of the Treasury, sums of such amounts as he may so determine for the purposes of this paragraph as being equivalent to the increase during that year of such liabilities of his as are attributable to the provision of pensions, allowances or gratuities to or in respect of persons who are or have been in the service of the Executive in so far as that increase results from the service of those persons during that accounting year and to the expense to be incurred in administering those pensions, allowances or gratuities.

Performance of functions

13. The Commission may authorise any member of the Commission or any officer or servant of the Commission or of the Executive to perform on behalf of the Commission such of the Commission's functions (including the function conferred on the Commission by this paragraph) as are specified in the authorisation.

Accounts and reports

14.–(1) It shall be the duty of the Commission–

   (a) to keep proper accounts and proper records in relation to the accounts;

   (b) to prepare in respect of each accounting year a statement of accounts in such form as the Secretary of State may direct with the approval of the Treasury; and

   (c) to send copies of the statement to the Secretary of State and the Comptroller and Auditor General before the end of the month of November next following the accounting year to which the statement relates.

(2) The Comptroller and Auditor General shall examine, certify and report on each statement received by him in pursuance of this Schedule and shall lay copies of each statement and of his report before each House of Parliament.

15. It shall be the duty of the Commission to make to the Secretary of State, as soon as possible after the end of each accounting year, a report on the performance of its functions during that year; and the Secretary of State shall lay before each House of Parliament a copy of each report made to him in pursuance of this paragraph.

Supplemental

16. The Secretary of State shall not make a determination or give his consent in pursuance of paragraph 5, 6, 7 or 11 of this Schedule except with the approval of the Minister for the Civil Service.

17. The fixing of the common seal of the Commission shall be authenticated by the signature of the secretary of the Commission or some other person authorised by the Commission to act for that purpose.

18. A document purporting to be duly executed under the seal of the Commission shall be received in evidence and shall, unless the contrary is proved, be deemed to be so executed.

19. In the preceding provisions of this Schedule-

    (a) "accounting year" means the period of twelve months ending with 31st March in any year except that the first accounting year of the Commission shall, if the Secretary of State so directs, be such period shorter or longer than twelve months (but not longer than two years) as is specified in the direction; and

    (b) "the chairman", "a deputy chairman" and "a member" mean respectively the chairman, a deputy chairman and a member of the Commission.

20.-(1) The preceding provisions of this Schedule (except paragraphs 10 to 12 and 15) shall have effect in relation to the Executive as if-

    (a) for any reference to the Commission there were substituted a reference to the Executive;

    (b) for any reference to the Secretary of State in paragraphs 2 to 4 and 19 and the first such reference in paragraph 7 there were substituted a reference to the Commission;

    (c) for any reference to the Secretary of State in paragraphs 5 to 7 (except the first such reference in paragraph 7) there were substituted a reference to the Commission acting with the consent of the Secretary of State;

    (d) for any reference to the chairman there were substituted a reference to the director and any reference to the deputy chairman were omitted;

    (e) in paragraph 14(1)(c) for the words from "Secretary" to "following" there were substituted the words "Commission by such date as the Commission may direct after the end of".

(2) It shall be the duty of the Commission to include in or send with the copies of the statement sent by it as required by paragraph 14(1)(c) of this Schedule copies of the statement sent to it by the Executive in pursuance of the said paragraph 14(1)(c) as adapted by the preceding sub-paragraph.

(3) The terms of an instrument appointing a person to be a member of the Executive shall be such as the Commission may determine with the approval of the Secretary of State and the Minister for the Civil Service.

SCHEDULE 3

Schedule 3 – *specifies the subject matter of health & safety regulations made under s15.*

1.–(1) Regulating or prohibiting-

(a) the manufacture, supply or use of any plant;

(b) the manufacture, supply, keeping or use of any substance;

(c) the carrying on of any process or the carrying out of any operation.

(2) Imposing requirements with respect to the design, construction, guarding, siting, installation, commissioning, examination, repair, maintenance, alteration, adjustment, dismantling, testing or inspection of any plant.

(3) Imposing requirements with respect to the marking of any plant or of any articles used or designed for use as components in any plant, and in that connection regulating or restricting the use of specified markings.

(4) Imposing requirements with respect to the testing, labelling or examination of any substance.

(5) Imposing requirements with respect to the carrying out of research in connection with any activity mentioned in sub-paragraphs (1) to (4) above.

2.–(1) Prohibiting the importation into the United Kingdom or the landing or unloading there of articles or substances of any specified description, whether absolutely or unless conditions imposed by or under the regulations are complied with.

(2) Specifying, in a case where an act or omission in relation to such an importation, landing or unloading as is mentioned in the preceding sub-paragraph constitutes an offence under a provision of this Act and of [the Customs and Excise Acts 1979], the Act under which the offence is to be punished.

3.–(1) Prohibiting or regulating the transport of articles or substances of any specified description.

(2) Imposing requirements with respect to the manner and means of transporting articles or substances of any specified description, including requirements with respect to the construction, testing and marking of containers and means of transport and the packaging and labelling of articles or substances in connection with their transport.

4.–(1) Prohibiting the carrying on of any specified activity or the doing of any specified thing except under the authority and in accordance with the terms and conditions of a licence, or except with the consent or approval of specified authority.

(2) Providing for the grant, renewal, variation, transfer and revocation of licences (including the variation and revocation of conditions attached to licences).

5. Requiring any person, premises or thing to be registered in any specified circumstances or as a condition of the carrying on of any specified activity or the doing of any specified thing.

6.–(1) Requiring, in specified circumstances, the appointment (whether in a specified capacity or not) of persons (or persons with specified qualifications or experience, or both) to perform specified functions, and imposing duties or conferring powers on persons appointed (whether in pursuance of the regulations or not) to perform specified functions.

(2) Restricting the performance of specified functions to persons possessing specified qualifications or experience.

7. Regulating or prohibiting the employment in specified circumstances of all persons or any class of persons.

8.–(1) Requiring the making of arrangements for securing the health of persons at work or other persons, including arrangements for medical examinations and health surveys.

(2) Requiring the making of arrangements for monitoring the atmospheric or other conditions in which persons work.

9. Imposing requirements with respect to any matter affecting the conditions in which persons work, including in particular such matters as the structural condition and stability of premises, the means of access to and egress from premises, cleanliness, temperature, lighting, ventilation, overcrowding, noise, vibrations, ionising and other radiations, dust and fumes.

10. Securing the provision of specified welfare facilities for persons at work, including in particular such things as an adequate water supply, sanitary conveniences, washing and bathing facilities, ambulance and first-aid arrangements, cloakroom accommodation, sitting facilities and refreshment facilities.

11. Imposing requirements with respect to the provision and use in specified circumstances of protective clothing or equipment, including affording protection against the weather.

12. Requiring in specified circumstances the taking of specified precautions in connection with the risk of fire.

13.-(1) Prohibiting or imposing requirements in connection with the emission into the atmosphere of any specified gas, smoke or dust or any other specified substance whatsoever.

(2) Prohibiting or imposing requirements in connection with the emission of noise, vibrations or any ionising or other radiations.

(3) Imposing requirements with respect to the monitoring of any such emission as is mentioned in the preceding sub-paragraphs.

14.-Imposing requirements with respect to the instruction, training and supervision of persons at work.

15.-(1) Requiring, in specified circumstances, specified matters to be notified in a specified manner to specified persons.

(2) Empowering inspectors in specified circumstances to require persons to submit written particulars of measures proposed to be taken to achieve compliance with any of the relevant statutory provisions.

16. Imposing requirements with respect to the keeping and preservation of records and other documents, including plans and maps.

17. Imposing requirements with respect to the management of animals.

18. The following purposes as regards premises of any specified description where persons work, namely–

(a) requiring precautions to be taken against dangers to which the premises or persons therein are or may be exposed by reason of conditions (including natural conditions) existing in the vicinity;

(b) securing that persons in the premises leave them in specified circumstances.

19. Conferring, in specified circumstances involving a risk of fire or explosion, power to search a person or any article which a person has with him for the purpose of ascertaining whether he has in his possession any article of a specified kind likely in those circumstances to cause a fire or explosion, and power to seize and dispose of any article of that kind found on such a search.

20. Restricting, prohibiting or requiring the doing of any specified thing where any accident or other occurrence of a specified kind has occurred.

21. As regards cases of any specified class, being a class such that the variety in the circumstances of particular cases within it calls for the making of special provision for particular cases, any of the following purposes, namely–

(a) conferring on employers or other persons power to make rules or give directions with respect to matters affecting health or safety;

(b) requiring employers or other persons to make rules with respect to any such matters;

(c) empowering specified persons to require employers or other persons either to make rules with respect to any such matters or to modify any such rules previously made by virtue of this paragraph; and

(d) making admissible in evidence without further proof, in such circumstances and subject to such conditions as may be specified, documents which purport to be copies of rules or rules of any specified class made under this paragraph.

22. Conferring on any local or public authority power to make byelaws with respect to any specified matter, specifying the authority or person by whom any byelaws made in the exercise of that power need to be confirmed, and generally providing for the procedure to be followed in connection with the making of any such byelaws.

Interpretation

23.–(1) In this Schedule "specified" means specified in health and safety regulations.

(2) It is hereby declared that the mention in this Schedule of a purpose that falls within any more general purpose mentioned therein is without prejudice to the generality of the more general purpose.

Schedule 4 – *contained modification of Part 1 of the Act relating to agriculture.*

Schedule 5 – *dealt with subject matter of Building regulations.*

Schedule 6 – *dealt with amendments of matters relating to Building regulations.*

Schedule 7 – *dealt with amendments to Building(Scotland)Act 1959.*

Now all repealed.

SCHEDULE 8

TRANSITIONAL PROVISIONS WITH RESPECT TO FIRE CERTIFICATES UNDER FACTORIES ACT 1961 OR OFFICES, SHOPS AND RAILWAY PREMISES ACT 1963

Schedule 8 – *deals with transitional provisions relating to fire certificates under Factories Act 1961 and Offices Shops and Railway Premises Act 1963 (both of the latter are now largely repealed)*

1. In this Schedule-

"the 1971 Act" means the Fire Precautions Act 1971;

"1971 Act certificate" means a fire certificate within the meaning of the 1971 Act;

"Factories Act certificate" means a certificate under section 40 of the Factories Act 1961 (means of escape in case of fire certification by fire authority);

"Office Act certificate" means a fire certification under section 29 of the Offices, Shops and Railway Premises Act 1963.

2.-(1) Where by virtue of an order under section 1 of the 1971 Act a 1971 Act certificate becomes required in respect of any premises at a time when there is in force in respect of those premises a Factories Act certificate or an Offices Act certificate ("the existing certificate"), the following provisions of this paragraph shall apply.

(2) The existing certificate shall continue in force (irrespective of whether the section under which it was issued remains in force) and–

(a) shall as from the said time be deemed to be a 1971 Act certificate validly issued with respect to the premises with respect to which it was issued and to cover the use or uses to which those premises were being put at that time; and

(b) may (in particular) be amended, replaced or revoked in accordance with the 1971 Act accordingly.

(3) Without prejudice to sub-paragraph (2)(b) above, the existing certificate, as it has effect by virtue of sub-paragraph (2) above, shall as from the said time be treated as imposing in relation to the premises the like requirements as were previously imposed in relation thereto by the following provisions, that is to say–

(a) if the existing certificate is a Factories Act certificate, the following provisions of the Factories Act 1961, namely sections 41(1), 48 (except subsections (5), (8) and (9)), 49(1), 51(1) and 52(1) and (4) and, so far as it relates to a proposed increase in the number of persons employed in any premises, section 41(3);

(b) if the existing certificate is an Offices Act certificate the following provisions of the Offices, Shops and Railways Act 1963, namely sections 30(1), 33, 34(1) and (2), 36(1) and 38(1) and, so far as it relates to a proposed increase in the number of persons employed to work in any premises at any one time, section 30(3).

3

Schedule 9 – *deals with minor and consequential amendments*
Schedule 10 – *dealt with repeals of other legislation.*
Now both repealed

The Management of Health and Safety at Work Regulations, 1999 No. 3242

Introductory Note

MHSWR 1992 were introduced in order to implement EU Framework Directive 89/391 plus part of the EU Temporary Workers Directive 91/383.

The 1992 MHSWR were replaced by MHSWR 1999 following changes made by 1994 and 1997 regulations in response to EU Pregnant Workers Directive and Protection of Young People at Work Directive 94/33 but most regulations from 1992 remain substantially unchanged. There is also an ACOP which employers may buy from HSE Books.92/85

MHSWR explains in more detail the requirements of HSW Act 1974. This is especially so concerning Health & safety Policies and risk assessments.

It is essential that management complies with these very important regulations as well as other relevant regulations (see Introduction to the Act for examples).

The Secretary of State, being a Minister designated (S.I. 1992/1711 and S.I. 1999/2027) for the purposes of section 2(2) of the European Communities Act 1972 (1972. c. 68; the enabling powers conferred by section 2(2) were extended by virtue of section 1 of the European Economic Area Act 1993 (c. 51)) in relation to measures relating to employers' obligations in respect of the health and safety of workers and in relation to measures relating to the minimum health and safety requirements for the workplace that relate to fire safety and in exercise of the powers conferred on him by the said section 2 and by sections 15(1), (2), (3)(a), (5), and (9), 47(2), 52(2), and (3), 80(1) and 82(3)(a) of and paragraphs 6(1), 7, 8(1), 10, 14, 15, and 16 of Schedule 3 to, the Health and Safety at Work etc. Act 1974 (1974 c. 37; sections 15 and 50 were amended by the Employment Protection Act 1975 (c. 71), Schedule 15, paragraphs 6 and 16 respectively) ("the 1974 Act") and of all other powers enabling him in that behalf—

    (a) for the purpose of giving effect without modifications to proposals submitted to him by the Health and Safety Commission under section 11(2)(d) of the 1974 Act after the carrying out by the Commission of consultations in accordance with section 50(3) of that Act; and

    (b) it appearing to him that the modifications to the Regulations marked with an asterisk in Schedule 2 are expedient and that it also appearing to him not to be appropriate to consult bodies in respect of such modifications in accordance with section 80(4) of the 1974 Act,

hereby makes the following Regulations:

Citation, commencement and interpretation

1.—(1) These Regulations may be cited as the Management of Health and Safety at Work Regulations 1999 and shall come into force on 29th December 1999.

(1)   MHSWR 1999 came into force on 29th December 1999. MHSWR 1992 came into effect on 1st January 1993.

(2)   In these Regulations—

"the 1996 Act" means the Employment Rights Act 1996 (1996 c. 18);

"the assessment" means, in the case of an employer or self-employed person, the assessment made or changed by him in accordance with regulation 3;

"child"—

(a) as respects England and Wales, means a person who is not over compulsory school age, construed in accordance with section 8 of the Education Act 1996 (1996 c. 56); and

(b) as respects Scotland, means a person who is not over school age, construed in accordance with section 31 of the Education (Scotland) Act 1980 (1980 c. 44);

"employment business" means a business (whether or not carried on with a view to profit and whether or not carried on in conjunction with any other business) which supplies persons (other than seafarers) who are employed in it to work for and under the control of other persons in any capacity;

"fixed-term contract of employment" means a contract of employment for a specific term which is fixed in advance or which can be ascertained in advance by reference to some relevant circumstance;

"given birth" means delivered a living child or, after twenty-four weeks of pregnancy, a stillborn child;

"new or expectant mother" means an employee who is pregnant; who has given birth within the previous six months; or who is breastfeeding;

"the preventive and protective measures" means the measures which have been identified by the employer or by the self-employed person in consequence of the assessment as the measures he needs to take to comply with the requirements and prohibitions imposed upon him by or under the relevant statutory provisions and by Part II of the Fire Precautions (Workplace) Regulations 1997 (S.I. 1997/1840; amended by S.I. 1999/1877);

"young person" means any person who has not attained the age of eighteen.

(2)   This sub paragraph contain definitions which apply throughout the Regulations. Most are self explanatory but additional explanation is given where necessary.

The ERA 1996 is the main statutory source of employment rights. It is now supplemented by the Employee Relations Act, 1999 and regulations made under both Acts. In Health and safety law there is some overlap with employment issues.

'the assessment' refers to 'risks assessments' made under regulation 3 whether modified or not.

'child' is someone not over 16 years of age (England and Wales).

'employment business' – includes employment agencies (profit or non profit making) who may supply agency staff to other organisations who may control those persons but who may NOT be their employers.

"fixed term contract'– e.g. One for 1 year, 2 years etc. and is not (which is more usual) an open ended contract terminable by notice.

'given birth' & 'new or expectant mother' – -these relate to pregnant workers health & safety rights (see regs. 17 & 18) & where this phrase is used only women who fit the definition gain protection.

NB. Pregnant workers also have employment rights (see ERA 1996).

'preventive and protective measures'–these are the measures that employers must implement following risk assessments in order to minimise risks to workers and others. They are sometimes referred to as 'control measures'. An employer must implement these under any regulations made under HSW Act 1974 (see list in Introduction to ACT) including the Fire Precautions (Workplace) Regulations,1997 which requires most employers to conduct a fire risk assessment.

(3)   Any reference in these Regulations to—
    (a) a numbered regulation or Schedule is a reference to the regulation or Schedule in these Regulations so numbered; or
    (b) a numbered paragraph is a reference to the paragraph so numbered in the regulation in which the reference appears.

(3)   fairly self explanatory. However, numbered paragraphs in Regulations are referred to as 'Regulation 1, 2 etc." in contrast to numbered paragraphs in Acts of Parliament which are referred to as 'sections' and 'sub sections'

Disapplication of these Regulations
2.—(1) These Regulations shall not apply to or in relation to the master or crew of a sea-going ship or to the employer of such persons in respect of the normal ship-board activities of a ship's crew under the direction of the master.

(1)   MHSWR 1999 is NOT applicable to employees on sea going ships or the employer in respect of normal ship board activities.

NB. There is separate Merchant Shipping legislation which is NOT as comprehensive as HSW Act 1974 & the regulations under it.

(2) Regulations 3(4), (5), 10(2) and 19 shall not apply to occasional work or short-term work involving—

    (a) domestic service in a private household; or

    (b) work regulated as not being harmful, damaging or dangerous to young people in a family undertaking.

> (2) This sub paragraph contains exemptions from protection provisions for young persons PROVIDED they are employed on an occasional basis or on short term work as a domestic servant in a private household (e.g. butlers maids etc.) or the work is in a family business where the work is not harmful, damaging or dangerous (e.g. helping in a family run shop).
>
> NB. Other legislation exists to protect children from exploitation in employment which employers must adhere to (even in family businesses) on pain of fines etc.

Risk assessment

> This is perhaps THE most important of all the Regulations and requires employers and self-employed to identify ALL significant hazards at work & conduct a risk assessment followed by preventive measures to protect others.
>
> Sometimes specific regulations will require a risk assessment (e.g. Health & safety (Display Screen Equipment) Regulations, 1992 or COSHH 1999) but where this is not the case or specific regulations contain exemptions, MHSWR 1999 reg. 3 requires a risk assessment for all significant risks.
>
> The ACOP to MHSWR 1999 (and other appropriate regulations) is a document employers should obtain as it contains much useful advice on implementation of the Regulations.

3.—(1) Every employer shall make a suitable and sufficient assessment of—

    (a) the risks to the health and safety of his employees to which they are exposed whilst they are at work; and

    (b) the risks to the health and safety of persons not in his employment arising out of or in connection with the conduct by him of his undertaking,

for the purpose of identifying the measures he needs to take to comply with the requirements and prohibitions imposed upon him by or under the relevant statutory provisions and by Part II of the Fire Precautions (Workplace) Regulations 1997.

> (1) Every employer must conduct a risk assessment of:
>
> (a) health & safety risks (inc. fire risks) to employees whilst at work: AND
>
> (b) health & safety risks (inc. fire risks) to other persons (non employees) who may be affected by the activities of the business. These people include; Neighbours, pedestrians, vehicle users, Contractors, contractor's employees,

visitors, those with a statutory right of entry, students, pupils, inmates, patients etc. and could in some circumstances include trespassers also. All these people are entitled to be protected.

The employer must then identify preventive measures to minimise the chance that an accident will occur.

The ACOP (an essential document available from HSE Books) explains what is a 'suitable and sufficient' risk assessment and its purpose and who should conduct it and reviews, recording, etc.

NB. Many large organisations will have devised their own risk assessment forms and procedures and employ in house expertise. Smaller organisations may gain some help from the very useful pamphlet, 'Five Steps to Risk Assessment' (and other publications) of HSE books. There are simple forms in this document which may be used or adapted by smaller businesses. If in doubt, a competent Health & safety Consultant should be called in to assist.

(2)   Every self-employed person shall make a suitable and sufficient assessment of—
    (a) the risks to his own health and safety to which he is exposed whilst he is at work; and
    (b) the risks to the health and safety of persons not in his employment arising out of or in connection with the conduct by him of his undertaking,

for the purpose of identifying the measures he needs to take to comply with the requirements and prohibitions imposed upon him by or under the relevant statutory provisions.

(2)   This sub paragraph requires self employed people (e.g. contractors) to conduct risk assessments concerning:
    (a) their own health & safety; and
    (b) others health & safety (broadly the same people as in 1 above with the addition of people who are NOT the self-employed person's own employees).

They too must implement preventive measures based on the risk assessment.

NB. The self-employed persons own employees are assessed under reg. 3(1) above. Under 2(a) if a self-employed person employs no one else, he must still conduct a risk assessment for his own health & safety.

(3)   Any assessment such as is referred to in paragraph (1) or (2) shall be reviewed by the employer or self-employed person who made it if—
    (a) there is reason to suspect that it is no longer valid; or
    (b) there has been a significant change in the matters to which it relates; and where as a result of any such review changes to an assessment are required, the employer or self-employed person concerned shall make them.

(3)   Risk Assessments in (1) and (2) above are NOT a once and for all activity. They must be reviewed frequently (e.g. once a year) but must in any case be reviewed if:

>   (a) it is out of date (e.g. because of new laws, advice, procedures 'best practice' etc. or the organisation is still having accidents and/or 'near misses' despite the original risk assessment.

>   (b) there are significant changes (e.g. changes to workplace layout or staff or procedures and processes or materials used etc).

If changes are required as a result of the review the employer must implement them.

(4) An employer shall not employ a young person unless he has, in relation to risks to the health and safety of young persons, made or reviewed an assessment in accordance with paragraphs (1) and (5).

> **(4)   It is illegal to employ a 'young person' (16-18) unless a risk assessment has been conducted under (1) above and (5) below. Prior risk assessments are important therefore if employing young persons.**

(5)   In making or reviewing the assessment, an employer who employs or is to employ a young person shall take particular account of—

>   (a) the inexperience, lack of awareness of risks and immaturity of young persons;
>   (b) the fitting-out and layout of the workplace and the workstation;
>   (c) the nature, degree and duration of exposure to physical, biological and chemical agents;
>   (d) the form, range, and use of work equipment and the way in which it is handled;
>   (e) the organisation of processes and activities;
>   (f) the extent of the health and safety training provided or to be provided to young persons; and
>   (g) risks from agents, processes and work listed in the Annex to Council Directive 94/33/EC (OJ No. L216, 20.8.94, p.12) on the protection of young people at work.

> **(5)   specific risk assessment (or review) in respect of employed 'young persons' must take account of paragraphs (a) to (g) which are fairly self explanatory. The purpose is to protect this most at risk group at work.**

(6)   Where the employer employs five or more employees, he shall record—
>   (a) the significant findings of the assessment; and
>   (b) any group of his employees identified by it as being especially at risk.

(6) Employers employing more than 5 workers must record:

(a) the significant findings of the risk assessment; and

(b) any group of employees identified as being especially at risk. (e.g. disabled persons, persons with susceptibility to harm, etc).

NB. The recording can be done on the HSE form (see above) or on one's own or on a PC (provided information is comprehensible and retrievable). The ACOP has further advice on this.

Employers with less than 5 are NOT exempted from doing risk assessment but merely from recording results. These employers are advised to still record in their own and their employees best interests.

Principles of prevention to be applied

4. Where an employer implements any preventive and protective measures he shall do so on the basis of the principles specified in Schedule 1 to these Regulations.

Following a risk assessment if preventive or control measures are needed in order to protect people, the employer must follow the hierarchy in Schedule 1 (see below) which provides a pattern for most regulations on Health & safety.

This means that ELIMINATION OF THE HAZARD is the first priority. If this cannot be done one moves down the list. Employers must NOT, for example, merely distribute personal protective equipment without first considering whether a measure higher up the hierarchy may protect people better.

Schedule 1 implements the provisions of Article 6(2) of EU Council Directive 89/391/EEC. This is a new 1999 regulation; it did not appear in the 1992 regulations.

NB. ALL SUBSEQUENT REGULATIONS HAVE BEEN RE-NUMBERED

(e.g. MHSWR 1992 reg. 4 is now MHSWR 1999 reg. 5 etc).

Health and safety arrangements

5.—(1) Every employer shall make and give effect to such arrangements as are appropriate, having regard to the nature of his activities and the size of his undertaking, for the effective planning, organisation, control, monitoring and review of the preventive and protective measures.

(1)   Every employer is under a duty to manage health & safety properly depending of the size and nature of his business and must pay particular attention to effective planning, organisation, control, monitoring and review of the preventive and protective measures identified by risk assessments.

This means that health and safety must NOT be left to chance but must be managed as with other considerations like profits, resources, staffing, product development, marketing etc.

The organisation's health & safety policy (required by s2(3) of HSW Act 1974) must contain statements regarding the health and safety responsibilities of everyone from the Chief Executive/Managing Director down to the lowest paid employee. Although 'employers' have the main responsibility many Companies designate the CE/MD as the person mainly responsible for health & safety in the organisation (see especially s37 of the Act). Departmental and line managers and supervisors should also be identified. Finally employees have duties (see s7/8 of the Act and reg. 14 of these regulations). These should all be spelled out.

In larger organisations 'Safety Officers or advisors or consultants' are employed to assist with these matters but the duties on others are not usually affected.

Many organisations compile an organisation chart and relate these duties to the levels on that chart. The purpose of all of this is so that HSE can identify those at fault if there are to be prosecutions for breach of the Act or any regulations under the Act.

Control measures following risk assessments must be implemented, monitored and reviewed to ensure their continued effectiveness and responsibilities for this allocated within the organisation as appropriate, perhaps with one person in a co-ordinatiing role.

NB. HSE Books can supply a draft Health & safety Policy form for use by smaller businesses. They also supply several publications on the Management of Risk which should be consulted and implemented by employers. If in doubt call in a competent health & safety consultant.

(2)   Where the employer employs five or more employees, he shall record the arrangements referred to in paragraph (1).

(2)   Where employers employ more than 5 workers the health & safety arrangements must be recorded (i.e. in the Health & safety Policy). Small employers are advised to record also as they are obliged to comply with reg. 5(1).

NB. The ACOP has further useful information on implementation.

Health surveillance

6. Every employer shall ensure that his employees are provided with such health surveillance as is appropriate having regard to the risks to their health and safety which are identified by the assessment.

Every employer if they have risks (identified by risk assessments) which can impair the health and safety of workers must conduct appropriate health surveillance.

This is repeated in several specific regulations.

E.G. Galvanisers who work with zinc are prone to cancer by the nature of their work. The employer must conduct medical checks on those employees at regular intervals at their own cost. The surveillance can be by an in house doctor or nurse, a GP or the NHS provided they are competent in the relevant field. With increasing stress at work an argument might be made that health surveillance should extend to this increasing hazard of work in view of recent cases. Audiometry tests should be conducted for those whose hearing is at risk at work.

Health Surveillance is highly appropriate for workers working with certain chemicals, asbestos, lead, ionising radiation etc.

NB. Further advice on implementation is contained in the ACOP and from EMAS.

Health and safety assistance

7.—(1) Every employer shall, subject to paragraphs (6) and (7), appoint one or more competent persons to assist him in undertaking the measures he needs to take to comply with the requirements and prohibitions imposed upon him by or under the relevant statutory provisions and by Part II of the Fire Precautions (Workplace) Regulations 1997.

(1)   This requires all employers (subject to exceptions in 6 & 7) below to appoint one or more 'competent persons' to assist him in implementing the requirements of all health & safety legislation (including risk assessments) and fire risk assessments under the Fire Precautions (Workplace) Regulations, 1997.

The ACOP defines 'competent persons' & makes it clear it does NOT necessarily depend on qualifications or particular skills (although the NEBOSH Certificate and/or Diploma & other qualifications, eg RSP, BSC, RoSPA, IOSH, NVQ's etc. are desirable as laws and procedures are complex and only those so trained may know them).Uncomplicated situations may only require someone:

(a) with an understanding of relevant current best practice;

(b) awareness of limitations of ones own experience & knowledge; and

(c) willingness & ability to supplement ones own experience and knowledge either by appointing a 'competent person' or employing a health & safety consultant.

The appointment of these persons does NOT absolve the employer of his responsibilities in law i.e. there can be delegation of tasks but NOT legal liability.

(see ACOP for more guidance)

(2)   Where an employer appoints persons in accordance with paragraph (1), he shall make arrangements for ensuring adequate co-operation between them.

(2)   Where two or more competent persons are appointed (e.g. separate people for COSHH and Electricity) there must be arrangements for ensuring co-ordination between them. A safety officer or advisor (or director with special responsibility and training) could be responsible for this.

(3)   The employer shall ensure that the number of persons appointed under paragraph (1), the time available for them to fulfil their functions and the means at their disposal are adequate having regard to the size of his undertaking, the risks to which his employees are exposed and the distribution of those risks throughout the undertaking.

(3)   The 'competent persons' shall be sufficient for the size of the organisation and time and resources must be made available to them to carry out proper risk assessments and control measures under reg. 3 to protect employees. Thus, a proper budget should be drawn up but one must be aware that emergencies can arise at any time and this budget may be exceeded. It is not a defence (esp. to a large organisation) that Employers have to await a new budget year.

(4) The employer shall ensure that—
    (a) any person appointed by him in accordance with paragraph (1) who is not in his employment—
        (i) is informed of the factors known by him to affect, or suspected by him of affecting, the health and safety of any other person who may be affected by the conduct of his undertaking, and
        (ii) has access to the information referred to in regulation 10; and

(4)   (a) The employer must ensure that any outside health & safety consultant who is an independent contractor—
        (i) is informed of any hazards etc. known or suspected in that organisation which are likely to affect the health & safety of any person, and
        (ii) the information listed in reg. 10 below.

Thus, outside consultants should NOT be expected to find their own way round. They must be informed of essential information about the undertaking in order that they can do their job properly.

The ACOP has further guidance as does an HSE pamphlet which advises on taking on consultants which is some help in avoiding the 'cowboys'.

    (b) any person appointed by him in accordance with paragraph (1) is given such information about any person working in his undertaking who is—
      (i) employed by him under a fixed-term contract of employment, or
      (ii) employed in an employment business,
    as is necessary to enable that person properly to carry out the function specified in that paragraph.

(4) (b) The employer must ensure that any other person appointed to be 'competent person' who is employed under a fixed term contract (e.g. of one year) or is recruited from an agency who employs him (i.e. agency staff) (as well as employees) are given all information necessary to enable them to carry out their functions (i.e. risk assessments and helping to implement health & safety laws and practice). Thus an employer cannot just appoint such persons and leave them to their own devices. They must take an active role and provide information and training also.

(5) A person shall be regarded as competent for the purposes of paragraphs (1) and (8) where he has sufficient training and experience or knowledge and other qualities to enable him properly to assist in undertaking the measures referred to in paragraph (1).

(5) This defines a 'competent person' as a person with sufficient training and experience or knowledge and other qualities to enable him properly to carry out his functions (see above).

Qualifications are not strictly necessary but with increasingly complex health & safety laws, procedures and practice, employers are advised to go for people with appropriate qualifications as well as the other factors.

(6) Paragraph (1) shall not apply to a self-employed employer who is not in partnership with any other person where he has sufficient training and experience or knowledge and other qualities properly to undertake the measures referred to in that paragraph himself.

(6) Self-employed sole traders are exempt from appointing 'competent persons' if they possess sufficient training and experience or knowledge and other qualities to enable him to carry out the functions of a 'competent person' himself.

E.G. An experienced grocer with few hazards in his shop may be competent to conduct simple risk assessments etc. relating to the generally less serious hazards he is likely to encounter. Self-employed electricians and gas fitters may well be 'competent persons' in their field.

NB. If they are NOT themselves 'competent persons' they must appoint and there is no exemption.

(7)   Paragraph (1) shall not apply to individuals who are employers and who are together carrying on business in partnership where at least one of the individuals concerned has sufficient training and experience or knowledge and other qualities—

(a) properly to undertake the measures he needs to take to comply with the requirements and prohibitions imposed upon him by or under the relevant statutory provisions; and

(b) properly to assist his fellow partners in undertaking the measures they need to take to comply with the requirements and prohibitions imposed upon them by or under the relevant statutory provisions.

(7)   Business partners are exempt from appointing 'competent persons' provided at least one of the partners has sufficient training and experience or knowledge and other qualities—

(a) properly to undertake risk assessments etc. in order to comply with health & safety laws and

(b) properly to assist his fellow partners to implement measures to comply with health & safety law.

NB. If they are NOT themselves 'competent persons' they must appoint and there is no exemption.

(8)   Where there is a competent person in the employer's employment, that person shall be appointed for the purposes of paragraph (1) in preference to a competent person not in his employment.

(8)   This is a new sub paragraph which was NOT in the 1992 regulations. Where 'competent persons' are already employed in the organisation as employees they shall be appointed in preference to outside health & safety consultants. The reason for this is that employees are likely to have a better 'feel' for the organisation and its layout, procedures etc. than outsiders. Where, however, an outsider is appointed because there is nobody in house, reg. 6(4) must be complied with.

Procedures for serious and imminent danger and for danger areas

8.—(1) Every employer shall—

    (a) establish and where necessary give effect to appropriate procedures to be followed in the event of serious and imminent danger to persons at work in his undertaking;

    (b) nominate a sufficient number of competent persons to implement those procedures in so far as they relate to the evacuation from premises of persons at work in his undertaking; and

    (c) ensure that none of his employees has access to any area occupied by him to which it is necessary to restrict access on grounds of health and safety unless the employee concerned has received adequate health and safety instruction.

(1) Every employer is bound by law to—

    (a) instigate procedures in respect of serious and imminent dangers to persons at work in his organisation;

    (b) appoint a sufficient number of 'competent persons' to implement evacuation procedures; and

    (c) exclude employees, other than those with appropriate health & safety training, from any sensitive areas at work.

The 'serious and imminent dangers' are not spelled out but include; fire, bombs, chemicals, radiation, etc. All organisations must have fire procedures based on fire risk assessments. Other dangers must be covered as appropriate (e.g. most government & local government buildings at risk of bomb hoaxes must have bomb evacuation procedures also).

'Persons at work' is a wider term than 'employees' and includes everyone. Thus, contractors, visitors etc. must be told what the evacuation procedures are. Many organisations do this by signing in visitors and giving them a copy of relevant procedures.

The 'competent persons' are different persons from those appointed under reg. 6 above. These are often called 'fire wardens', 'fire marshals', 'sweepers' etc.

Sensitive areas at work include chemical, radiation areas etc. and they should be locked off and 'permit to work' systems established overseen by supervisors and backed up by appropriate training for all concerned.

(2) Without prejudice to the generality of paragraph (1)(a), the procedures referred to in that sub-paragraph shall—

    (a) so far as is practicable, require any persons at work who are exposed to serious and imminent danger to be informed of the nature of the hazard and of the steps taken or to be taken to protect them from it;

(b) enable the persons concerned (if necessary by taking appropriate steps in the absence of guidance or instruction and in the light of their knowledge and the technical means at their disposal) to stop work and immediately proceed to a place of safety in the event of their being exposed to serious, imminent and unavoidable danger; and

(c) save in exceptional cases for reasons duly substantiated (which cases and reasons shall be specified in those procedures), require the persons concerned to be prevented from resuming work in any situation where there is still a serious and imminent danger.

(2) Without diluting the duties in (1)(a) above the evacuation procedures must–

(a) inform people at risk of the nature of the hazards and preventive measures, so far as is reasonably practicable;

(b) enable persons at risk (on their own initiative if necessary) to stop work immediately and go to a place of safety which protects them from the danger

(c) prevent people from resuming work (except in exceptional cases where substantial reasons exist) where there is still a serious and imminent danger.

As well as notices, (a) – (c) above should be complied with by building into induction training sessions on these procedures plus updates. Contractors and visitors passes should give information on these procedures.

(3) A person shall be regarded as competent for the purposes of paragraph (1)(b) where he has sufficient training and experience or knowledge and other qualities to enable him properly to implement the evacuation procedures referred to in that sub-paragraph.

(3) This definition is the same as for Reg.7(5) above BUT separate persons may need to be appointed for each area/floor of the workplace.

NB. The ACOP contains further guidance.

Contacts with external services

9. Every employer shall ensure that any necessary contacts with external services are arranged, particularly as regards first-aid, emergency medical care and rescue work.

This is a new 1999 regulation which did not appear in the 1992 regulations.

Every employer is required by law to set up procedures and persons responsible for phoning emergency services such as fire brigade, police, ambulance, hospitals and other experts e.g. air sea rescue (if appropriate). This should NOT be left to chance; it is too late in an emergency situation and people should be designated to contact the appropriate services.

Information for employees

10.—(1) Every employer shall provide his employees with comprehensible and relevant information on—

(a) the risks to their health and safety identified by the assessment;

(b) the preventive and protective measures;

(c) the procedures referred to in regulation 8(1)(a) and the measures referred to in regulation 4(2)(a) of the Fire Precautions (Workplace) Regulations 1997;

(d) the identity of those persons nominated by him in accordance with regulation 8(1)(b) and regulation 4(2)(b) of the Fire Precautions (Workplace) Regulations 1997; and

(e) the risks notified to him in accordance with regulation 11(1)(c).

> (1) Every employer must by law provide his employees with understandable and relevant information on–
>
> - risk assessments
> - preventive/control measures
> - emergency evacuation procedures(inc. fire evacuation)
> - identity of 'competent persons' to oversee evacuations
> - information on risks informed to the employer by other employers
>
> This information should be available in a retrievable & accessible form at all times whilst persons are at work (inc. evening shifts) and should be incorporated into induction training and staff handbooks, noticeboards etc.

(2) Every employer shall, before employing a child, provide a parent of the child with comprehensible and relevant information on—

(a) the risks to his health and safety identified by the assessment;

(b) the preventive and protective measures; and

(c) the risks notified to him in accordance with regulation 11(1)(c).

> (2) Employers must, before employing a child provide the parent of that child with understandable and relevant information on;
>
> - risk assessments
> - preventive/control measures
> - evacuation procedures

(3) The reference in paragraph (2) to a parent of the child includes—

(a) in England and Wales, a person who has parental responsibility, within the meaning of section 3 of the Children Act 1989 (1989 c. 41), for him; and

(b) in Scotland, a person who has parental rights, within the meaning of section

8 of the Law Reform (Parent and Child) (Scotland) Act 1986 (1986 c. 9) for him.

> (3) 'Child' is defined in s3 Children Act 1989 (England & Wales) & s8 Law Reform (Parent & Child) (Scotland) Act,1986.
>
> NB. The ACOP contains further guidance.

Co-operation and co-ordination

11.—(1) Where two or more employers share a workplace (whether on a temporary or a permanent basis) each such employer shall—

    (a) co-operate with the other employers concerned so far as is necessary to enable them to comply with the requirements and prohibitions imposed upon them by or under the relevant statutory provisions and by Part II of the Fire Precautions (Workplace) Regulations 1997;

    (b) (taking into account the nature of his activities) take all reasonable steps to co-ordinate the measures he takes to comply with the requirements and prohibitions imposed upon him by or under the relevant statutory provisions and by Part II of the Fire Precautions (Workplace) Regulations 1997 with the measures the other employers concerned are taking to comply with the requirements and prohibitions imposed upon them by that legislation; and

    (c) take all reasonable steps to inform the other employers concerned of the risks to their employees' health and safety arising out of or in connection with the conduct by him of his undertaking.

> (1) Where more than one employer shares a workplace with another or others (whether temporarily or permanently) each employer has a legal duty to—
>
>   (a) co-operate with the other employers so far as is necessary to enable each to comply with health & safety laws and fire prevention laws.
>
>   (b) co-ordinate activities with other employers (taking account of risks in his business) to comply with health & safety and fire prevention laws.
>
>   (c) inform other employers of risks in his undertaking which may affect other employees.
>
> For example, if several organisations/employers share an office block they must obviously co-operate & co-ordinate their activities concerning fire evacuation, lift safety, common stairways, halls, passages etc. and additionally inform all the other employers of specific risks that may affect them.
>
> This is to ensure that each is not "doing their own thing" which may be dangerous in an emergency.

(2)   Paragraph (1) (except in so far as it refers to Part II of the Fire Precautions (Workplace) Regulations 1997) shall apply to employers sharing a workplace with self-employed persons and to self-employed persons sharing a workplace with other self-employed persons as it applies to employers sharing a workplace with other employers; and the references in that paragraph to employers and the reference in the said paragraph to their employees shall be construed accordingly.

> (2) (1) applies to:
>
> - employers sharing with other employers
>
> - employers sharing with self employed persons
>
> - self employed persons sharing with self employed persons.
>
> Surprisingly, fire precautions laws do NOT relate to the last two groups BUT it is sensible still to comply for it may be negligent in civil law not to.
>
> NB. The ACOP contains further guidance.

Persons working in host employers' or self-employed persons' undertakings
12.—(1) Every employer and every self-employed person shall ensure that the employer of any employees from an outside undertaking who are working in his undertaking is provided with comprehensible information on—
   (a) the risks to those employees' health and safety arising out of or in connection with the conduct by that first-mentioned employer or by that self-employed person of his undertaking; and
   (b) the measures taken by that first-mentioned employer or by that self-employed person in compliance with the requirements and prohibitions imposed upon him by or under the relevant statutory provisions and by Part II of the Fire Precautions (Workplace) Regulations 1997 in so far as the said requirements and prohibitions relate to those employees.

> (1)   Every employer and self-employed person must ensure that employers of other people in their premises are provided with understable information on:
>
> (a) risk assessments to those other employees regarding risks in his organisation to which they may be subject
>
> (b) preventive and control measures to minimise the health & safety risks at (a) above including fire risks.
>
> Thus, all contractors must be provided with appropriate information regarding their employees health & safety and should ensure that that information is passed on their employees. This should be sewn up in contracts also.

(2)   Paragraph (1) (except in so far as it refers to Part II of the Fire Precautions (Workplace) Regulations 1997) shall apply to a self-employed person who is working in the undertaking of an employer or a self-employed person as it applies to employees from an outside undertaking who are working therein; and the reference in that paragraph to the employer of any employees from an outside undertaking who are working in the undertaking of an employer or a self-employed person and the references in the said paragraph to employees from an outside undertaking who are working in the undertaking of an employer or a self-employed person shall be construed accordingly.

(2) (1) above also refers to

- self-employed working in an employers organisation
- self-employed working in another self-employed persons organisation.

They must be treated as 'employees' and given the same information, except for fire risks, surprisingly. It is best to include this also for the reasons stated above.

(3)   Every employer shall ensure that any person working in his undertaking who is not his employee and every self-employed person (not being an employer) shall ensure that any person working in his undertaking is provided with appropriate instructions and comprehensible information regarding any risks to that person's health and safety which arise out of the conduct by that employer or self-employed person of his undertaking.

(3)   All employers must ensure that all contractors employees and lone self-employed persons working in his premises are provided with appropriate instructions and understandable information regarding health & safety risks they may be subject to.

(4)   Every employer shall—
   (a) ensure that the employer of any employees from an outside undertaking who are working in his undertaking is provided with sufficient information to enable that second-mentioned employer to identify any person nominated by that first mentioned employer in accordance with regulation 8(1)(b) to implement evacuation procedures as far as those employees are concerned; and
   (b) take all reasonable steps to ensure that any employees from an outside undertaking who are working in his undertaking receive sufficient information to enable them to identify any person nominated by him in accordance with regulation 8(1)(b) to implement evacuation procedures as far as they are concerned.

(4) All employers must—

(a) ensure that contractors working in his premises are provided with sufficient information in order to identify 'competent persons' responsible for evacuation procedures under reg. 8 (1)(b); and

(b) take reasonable steps to ensure that contractors employees receive the same information as in (a) above.

(5) Paragraph (4) shall apply to a self-employed person who is working in an employer's undertaking as it applies to employees from an outside undertaking who are working therein; and the reference in that paragraph to the employer of any employees from an outside undertaking who are working in an employer's undertaking and the references in the said paragraph to employees from an outside undertaking who are working in an employer's undertaking shall be construed accordingly.

(5) (4) applies also to self-employed persons working in employers premises and they are treated as 'employees' for this purpose (i.e. the same information must be given to them).

NB. The ACOP contains further guidance. Employers are advised to establish 'Contractors procedures' to cover the above and other matters relating to contractors (e.g. CDM regs. and contractual matters).

Capabilities and training

13.—(1) Every employer shall, in entrusting tasks to his employees, take into account their capabilities as regards health and safety.

(1) When employers employ staff they must not only be concerned with job skills but also health and safety factors for the job including capability. Training should be given as necessary for all jobs. This reg. reinforces s2(2)(c) of HSW Act 1974 (see above)

(2) Every employer shall ensure that his employees are provided with adequate health and safety training—

(a) on their being recruited into the employer's undertaking; and

(b) on their being exposed to new or increased risks because of—

(i) their being transferred or given a change of responsibilities within the employer's undertaking,

(ii) the introduction of new work equipment into or a change respecting work equipment already in use within the employer's undertaking,

(iii) the introduction of new technology into the employer's undertaking, or

(iv) the introduction of a new system of work into or a change respecting a system of work already in use within the employer's undertaking.

(2)   All employers must ensure that all employees are provided with adequate health & safety training—

(a) on first recruitment to the job (i.e. induction training); and

(b) on being exposed to new or increased risks because of—

(i) transfers or changes of job responsibilities

(ii) introduction of new work equipment or changes in work equipment

(iii) introduction of new technology (e.g. display screen equipment etc.), or

(iv) introduction of new systems of work or changes to systems of work.

Those given health & safety responsibilities, for example, must be appropriately trained.

(3) The training referred to in paragraph (2) shall—
   (a) be repeated periodically where appropriate;
   (b) be adapted to take account of any new or changed risks to the health and safety of the employees concerned; and
   (c) take place during working hours.

(3)   The training at (2) must—

(a) be repeated periodically as appropriate (i.e. refresher training);

(b) be adapted to take account of new or changed risks to employees; and

(c) take place during working hours. No loss of pay or other benefits should result.

NB. The ACOP gives further guidance on these matters.

Employees' duties

This reg. expands on duties contained in ss7 & 8 HSW Act 1974 and other health & safety regulations.

14.—(1) Every employee shall use any machinery, equipment, dangerous substance, transport equipment, means of production or safety device provided to him by his employer in accordance both with any training in the use of the equipment concerned which has been received by him and the instructions respecting that use which have been provided to him by the said employer in compliance with the requirements and prohibitions imposed upon that employer by or under the relevant statutory provisions.

> **(1)** All employees must use work equipment, chemicals, transport, means of production and safety devices provided by the employer in accordance with training and instruction given in compliance with health & safety law (inc. s2(2)(c) HSW Act 1974)

(2) Every employee shall inform his employer or any other employee of that employer with specific responsibility for the health and safety of his fellow employees—

(a) of any work situation which a person with the first-mentioned employee's training and instruction would reasonably consider represented a serious and immediate danger to health and safety; and

(b) of any matter which a person with the first-mentioned employee's training and instruction would reasonably consider represented a shortcoming in the employer's protection arrangements for health and safety,

in so far as that situation or matter either affects the health and safety of that first mentioned employee or arises out of or in connection with his own activities at work, and has not previously been reported to his employer or to any other employee of that employer in accordance with this paragraph.

> **(2)** All employees must inform the employer or person entrusted with health & safety after duties—
>
> (a) of any situation at work which represents a serious and immediate danger to health & safety, taking into account the employees training and instruction; and
>
> (b) of any shortcomings in the employers health & safety protection arrangements.
>
> Thus, all employees are safety inspectors to the extent that they have been trained and instructed to recognise problems.
>
> Employees can be fined up to 5000 pounds in a Magistrates' Court for breach unless they have a defence because they lack the training or instruction from the employer. It is therefore in the employer's interests to conduct that training and instruction and to set up sensible procedures for reporting. The employer does not escape his responsibility to comply with statutory duties.
>
> NB. The ACOP contains useful guidance on implementation.

Temporary workers

15.—(1) Every employer shall provide any person whom he has employed under a fixed-term contract of employment with comprehensible information on—

(a) any special occupational qualifications or skills required to be held by that employee if he is to carry out his work safely; and

(b) any health surveillance required to be provided to that employee by or under any of the relevant statutory provisions,

and shall provide the said information before the employee concerned commences his duties.

> **(1)** Employers who employ people on fixed term contracts (e.g. 1 year long) must provide these temporary employees with understandable information prior to their commencing employment on—
>
> **(a)** any special occupational qualifications or skills in order that the job may be carried out safely; and
>
> **(b)** any appropriate health surveillance under any health and safety legislation.
>
> This is an extra safeguard for temporary workers in addition to their right to other information under s2(2)(c) of HSW Act 1974 and other regulations.

(2) Every employer and every self-employed person shall provide any person employed in an employment business who is to carry out work in his undertaking with comprehensible information on—

(a) any special occupational qualifications or skills required to be held by that employee if he is to carry out his work safely; and

(b) health surveillance required to be provided to that employee by or under any of the relevant statutory provisions.

> **(2)** Employers and self-employed persons who are to take on agency staff must provide the same prior information as in (1) above to the worker.

(3) Every employer and every self-employed person shall ensure that every person carrying on an employment business whose employees are to carry out work in his undertaking is provided with comprehensible information on—

(a) any special occupational qualifications or skills required to be held by those employees if they are to carry out their work safely; and

(b) the specific features of the jobs to be filled by those employees (in so far as those features are likely to affect their health and safety);

and the person carrying on the employment business concerned shall ensure that the information so provided is given to the said employees.

> **(3)** Employers and self-employed persons must provide the Employment Agency with the same information as in (1) and the Employment agency must ensure it is given to their employees.

Risk assessment in respect of new or expectant mothers

16.—(1) Where—

    (a) the persons working in an undertaking include women of child-bearing age; and

    (b) the work is of a kind which could involve risk, by reason of her condition, to the health and safety of a new or expectant mother, or to that of her baby, from any processes or working conditions, or physical, biological or chemical agents, including those specified in Annexes I and II of Council Directive 92/85/EEC (OJ No. L348, 28.11.92, p.1) on the introduction of measures to encourage improvements in the safety and health at work of pregnant workers and workers who have recently given birth or are breastfeeding,

the assessment required by regulation 3(1) shall also include an assessment of such risk.

This is a new regulation as a result of EU legislation.

(1) Where—

    (a) employees include women of child bearing age; and

    (b) the nature of the work is such that there could be extra risks to the health and safety of new or expectant mothers or their babies arising from processes, working conditions, physical, biological or chemical agents (inc. those listed in Annexes I & II of EU Council Directive 92/85/EEC), a specific risk assessment for such persons shall be done under reg3(1).

NB. Examples of risks as well as the above could include work on DSE or manual handling hazards, radiation, etc.

(2) Where, in the case of an individual employee, the taking of any other action the employer is required to take under the relevant statutory provisions would not avoid the risk referred to in paragraph (1) the employer shall, if it is reasonable to do so, and would avoid such risks, alter her working conditions or hours of work.

(2) Where in respect of an individual mother general preventive measures would not protect her the employer must (if it is reasonable to do so and it would avoid risks) alter her working conditions or hours.

This means that options are, to reduce hours or give special leave or re-deploy to other work during pregnancy and for a while after birth.

If in any doubt at all good employers should take pregnant employees off potentially dangerous work without loss of pay or conditions.

(3)   If it is not reasonable to alter the working conditions or hours of work, or if it would not avoid such risk, the employer shall, subject to section 67 of the 1996 Act suspend the employee from work for so long as is necessary to avoid such risk.

> **(3)   If it is not reasonable to alter the pregnant employees working conditions or hours, the employer must suspend the employee for as long as is necessary to avoid risks, subject to s67 ERA 1996.**

(4)   In paragraphs (1) to (3) references to risk, in relation to risk from any infectious or contagious disease, are references to a level of risk at work which is in addition to the level to which a new or expectant mother may be expected to be exposed outside the workplace.

> **(4)   In (1) to (3) above 'risk' in relation to infectious or contagious diseases, relates to levels of risk in the workplace above those which a new or expectant mother would be exposed to outside the workplace.**
>
> **Thus, a common cold may not qualify but legionella as a result of faulty cooling systems might.**
>
> **Risk assessment forms should allow therefore for special risk assessments of pregnant women and nursing mothers.**
>
> **NB. See the ACOP for further guidance.**

Certificate from registered medical practitioner in respect of new or expectant mothers

17.   Where—
   (a) a new or expectant mother works at night; and
   (b) a certificate from a registered medical practitioner or a registered midwife shows that it is necessary for her health or safety that she should not be at work for any period of such work identified in the certificate,
   the employer shall, subject to section 67 of the 1996 Act, suspend her from work for so long as is necessary for her health or safety.

> **Where—**
>   **(a) a new or expectant mother works at night; and**
>   **(b) a GP or midwife issues a certificate that she should not work for any period indicated in the certificate, the employer must suspend the employee for as long as is necessary for her health & safety, subject to s67 ERA 1996.**
>
> **Redeployment etc. need not be considered in such cases it seems.**

Notification by new or expectant mothers

18.—(1) Nothing in paragraph (2) or (3) of regulation 16 shall require the employer to take any action in relation to an employee until she has notified the employer in writing that she is pregnant, has given birth within the previous six months, or is breastfeeding.

> (1) The rights in reg. 16(2) and (3) do NOT come into play until the worker has notified the employer in writing that she is pregnant, has given birth within the previous 6 months or is breastfeeding. The onus is thus on the employee.

(2) Nothing in paragraph (2) or (3) of regulation 16 or in regulation 17 shall require the employer to maintain action taken in relation to an employee—

    (a) in a case—

        (i) to which regulation 16(2) or (3) relates; and

        (ii) where the employee has notified her employer that she is pregnant, where she has failed, within a reasonable time of being requested to do so in writing by her employer, to produce for the employer's inspection a certificate from a registered medical practitioner or a registered midwife showing that she is pregnant;

    (b) once the employer knows that she is no longer a new or expectant mother; or

    (c) if the employer cannot establish whether she remains a new or expectant mother.

> (2) An employer need not maintain special action under regs. 16(2)(3) or reg. 17 to protect a pregnant worker—
>
>   (a) in cases to which reg. 16(2) or (3) apply where the pregnant employee has notified the pregnancy but has failed within a reasonable time to provide medical certificates for that;
>
>   (b) once the employer knows she is no longer a new or expectant mother; or
>
>   (c) if the employer cannot establish whether she remains a new or expectant mother.
>
> The protective provisions are not automatically tripped therefore, they depend on notification and continued notification.
>
> NB. Employers could, however, develop forms which employees could use for this purpose if they so choose.
>
> See the ACOP for further guidance.

Protection of young persons

19.—(1) Every employer shall ensure that young persons employed by him are protected at work from any risks to their health or safety which are a consequence of their lack of experience, or absence of awareness of existing or potential risks or the fact that young persons have not yet fully matured.

> This is a new 1999 regulation effective from 29th December 1999.
>
> (1)   It is mandatory on every employer to ensure that employed young persons are protected at work from any health and safety risks arising from the young person's inexperience, lack of perception of risks or immaturity. Thus, special training, supervision, information etc. must be given It may well be that certain work equipment and systems should NOT be operated by young persons.

(2) Subject to paragraph (3), no employer shall employ a young person for work—
   (a) which is beyond his physical or psychological capacity;
   (b) involving harmful exposure to agents which are toxic or carcinogenic, cause heritable genetic damage or harm to the unborn child or which in any other way chronically affect human health;
   (c) involving harmful exposure to radiation;
   (d) involving the risk of accidents which it may reasonably be assumed cannot be recognised or avoided by young persons owing to their insufficient attention to safety or lack of experience or training; or
   (e) in which there is a risk to health from—
      (i)   extreme cold or heat;
      (ii) noise; or
      (iii) vibration,

and in determining whether work will involve harm or risks for the purposes of this paragraph, regard shall be had to the results of the assessment.

> (2)   This bans employers employing young persons for work
>
>   (a) which exceeds their physical or psychological capabilities (e.g activities mature adults may be able to do, young persons cannot)
>
>   (b) where they would be exposed to harmful poisonous or cancer inducing agents (e.g. chemicals etc.), or cause damage which may be passed on in the genes to future generations or harm a foetus or which in any other way affects health on a long term basis
>
>   (c) where they would be exposed to harmful radiation;
>
>   (d) where they would be at risk of accidents which they may not be able to recognise owing to insufficient attention span, lack of experience or training; or
>
>   (e) where there would be risks to health from extremes of temperature, noise or vibration (e.g. hand held power tools etc.).
>
> In assessing (a) to (e) above regard must be had to the risk assessment.

(3)   Nothing in paragraph (2) shall prevent the employment of a young person who is no longer a child for work—

(a) where it is necessary for his training;

(b) where the young person will be supervised by a competent person; and

(c) where any risk will be reduced to the lowest level that is reasonably practicable.

(3) Despite reg. 19(2) an employer is NOT prevented from employing a young person (who is no longer a child) for work

(a) where it is necessary for his training

(b) where there is supervision by a competent person (see reg. 7 definition above), or

(c) where risks are reduced to the lowest level reasonably practicable.

(4) The provisions contained in this regulation are without prejudice to—

(a) the provisions contained elsewhere in these Regulations; and

(b) any prohibition or restriction, arising otherwise than by this regulation, on the employment of any person.

(4) Provisions in reg. 19 are without prejudice to—

(a) other provisions of these regs. (e.g. risk assessment under reg. 3)

(b) prohibitions or restrictions on employment of any person contained in any other Statutes or regulations (e.g. restrictive hours of work for children in other legislation).

Exemption certificates

20.—(1) The Secretary of State for Defence may, in the interests of national security, by a certificate in writing exempt—

(a) any of the home forces, any visiting force or any headquarters from those requirements of these Regulations which impose obligations other than those in regulations 16-18 on employers; or

(b) any member of the home forces, any member of a visiting force or any member of a headquarters from the requirements imposed by regulation 14;

and any exemption such as is specified in sub-paragraph (a) or (b) of this paragraph may be granted subject to conditions and to a limit of time and may be revoked by the said Secretary of State by a further certificate in writing at any time.

(2) In this regulation—

(a) "the home forces" has the same meaning as in section 12(1) of the Visiting Forces Act 1952 (1952 c. 67);

(b) "headquarters" means a headquarters for the time being specified in Schedule 2 to the Visiting Forces and International Headquarters

(Application of Law) Order 1999 (S.I. 1999/1736);

(c) "member of a headquarters" has the same meaning as in paragraph 1(1) of the Schedule to the International Headquarters and Defence Organisations Act 1964 (1964 c.5); and

(d) "visiting force" has the same meaning as it does for the purposes of any provision of Part I of the Visiting Forces Act 1952.

> This regulation relates to home or foreign forces who may seek an exemption certificate from certain regulations in the interests of national security from the ministry of defence. Forces may be exempted from all regs. except Regs. 16-18. Individual forces members may be exempted from reg. 14.
>
> Certificates may be subject to conditions or limited as to time.

Provisions as to liability

21. Nothing in the relevant statutory provisions shall operate so as to afford an employer a defence in any criminal proceedings for a contravention of those provisions by reason of any act or default of—

(a) an employee of his, or

(b) a person appointed by him under regulation 7.

> An employer may not as a defence to any criminal proceedings under any health & safety legislation plead that the act was the fault of an employee of his or a 'competent person' appointed under reg. 7.
>
> The employer AND these people may be prosecuted if all share the blame. An employer cannot legally delegate his duties in any case.

Exclusion of civil liability

22.—(1) Breach of a duty imposed by these Regulations shall not confer a right of action in any civil proceedings.

> (1) Breach of these regulations give rise to purely criminal proceedings and do NOT (as with some other regulations) give rise also to a breach of statutory duty action at civil law.

(2) Paragraph (1) shall not apply to any duty imposed by these Regulations on an employer—

(a) to the extent that it relates to risk referred to in regulation 16(1) to an employee; or

(b) which is contained in regulation 19.

> (2) There are two exceptions to (1) which allow pregnant women and nursing mothers and young persons at work to sue for breach of statutory duty if special risk assessments have not been carried out to protect them, subject to other conditions relating to the bringing of this civil action.
>
> NB. Subject to the exceptions at (2) above MHSWR 1999 are purely criminal law regulations.

Extension outside Great Britain

23.—(1) These Regulations shall, subject to regulation 2, apply to and in relation to the premises and activities outside Great Britain to which sections 1 to 59 and 80 to 82 of the Health and Safety at Work etc. Act 1974 apply by virtue of the Health and Safety at Work etc. Act 1974 (Application Outside Great Britain) Order 1995 (S.I. 1995/263) as they apply within Great Britain.

> (1) These regulations (subject to reg. 2) apply to offshore oil and gas rigs etc. in relation to the stated sections of the HSW Act 1974.

(2) For the purposes of Part I of the 1974 Act, the meaning of "at work" shall be extended so that an employee or a self-employed person shall be treated as being at work throughout the time that he is present at the premises to and in relation to which these Regulations apply by virtue of paragraph (1); and, in that connection, these Regulations shall have effect subject to the extension effected by this paragraph.

> (2) In Part 1 of HSW Act 1974 the definition of "at work" is extended so that employees or self employed persons are treated as being at work at all times they are present on oil or gas rigs(even though they might be resting).

Amendment of the Health and Safety (First-Aid) Regulations 1981

24. Regulation 6 of the Health and Safety (First-Aid) Regulations 1981 (S.I. 1981/917; amended by S.I. 1989/1671) is hereby revoked.

Amendment of the Offshore Installations and Pipeline Works (First-Aid) Regulations 1989

25.—(1) The Offshore Installations and Pipeline Works (First-Aid) Regulations 1989 (S.I. 1989/1671; amended by S.I. 1993/1823, and S.I. 1995/738) shall be amended in accordance with the following provisions of this regulation.

(2) In regulation 7(1) for the words "from all or any of the requirements of these Regulations", there shall be substituted the words "from regulation 5(1)(b) and (c) and (2)(a) of these Regulations".

(3) After regulation 7(2) the following paragraph shall be added—

"(3) An exemption granted under paragraph (1) above from the requirements in regulation 5(2)(a) of these Regulations shall be subject to the condition that a person provided under regulation 5(1)(a) of these Regulations shall have undergone adequate training.".

Amendment of the Mines Miscellaneous Health and Safety Provisions Regulations 1995

26.—(1) The Mines Miscellaneous Health and Safety Provisions Regulations 1995 (S.I. 1995/2005) shall be amended in accordance with the following provisions of this regulation.

(2) Paragraph (2)(b) of regulation 4 shall be deleted.

(3) After paragraph (4) of regulation 4 there shall be added the following paragraph—

"(5) In relation to fire, the health and safety document prepared pursuant to paragraph (1) shall—

(a) include a fire protection plan detailing the likely sources of fire, and the precautions to be taken to protect against, to detect and combat the outbreak and spread of fire; and

(b) in respect of every part of the mine other than any building on the surface of that mine—

    (i) include the designation of persons to implement the plan, ensuring that the number of such persons, their training and the equipment available to them is adequate, taking into account the size of, and the specific hazards involved in the mine concerned; and

    (ii) include the arrangements for any necessary contacts with external emergency services, particularly as regards rescue work and fire-fighting; and

    (iii) be adapted to the nature of the activities carried on at that mine, the size of the mine and take account of the persons other than employees who may be present.".

Amendment of the Construction (Health, Safety and Welfare) Regulations 1996

27.—(1) The Construction (Health, Safety and Welfare) Regulations 1996 (S.I. 1996/1592) shall be amended in accordance with the following provisions of this regulation.

(2) Paragraph (2) of regulation 20 shall be deleted and the following substituted—

"(2) Without prejudice to the generality of paragraph (1), arrangements prepared pursuant to that paragraph shall—

(a) have regard to those matters set out in paragraph (4) of regulation 19;

(b) designate an adequate number of persons who will implement the arrangements; and

(c) include any necessary contacts with external emergency services, particularly as regards rescue work and fire-fighting.".

These regulations revoke or amend other regulations relating to first aid (onshore and offshore), Mines and Construction, the scope of which are outside the remit of this book.

Regulations to have effect as health and safety regulations

28. Subject to regulation 9 of the Fire Precautions (Workplace) Regulations 1997 (S.I. 1997/1840; amended by S.I. 1999/1877), these Regulations shall, to the extent that they would not otherwise do so, have effect as if they were health and safety regulations within the meaning of Part I of the Health and Safety at Work etc. Act 1974.

Certain regulations although not made under s15 HSW Act 1974 are treated as if they were so made despite being made under other legislation. The Fire Precautions (Workplace) Regulations 1997 are thus included which permits the risk assessment requirements of these regs (see reg. 3)to be applied to them.

Revocations and consequential amendments

29.—(1) The Management of Health and Safety at Work Regulations 1992 (S.I. 1992/2051; amended by S.I. 1994/2865; S.I. 1997/135, and S.I. 1997/1840), the Management of Health and Safety at Work (Amendment) Regulations 1994 (S.I. 1994/2865), the Health and Safety (Young Persons) Regulations 1997 (S.I. 1997/135) and Part III of the Fire Precautions (Workplace) Regulations 1997 are hereby revoked.

(1) Older Management regs. and others are repealed. However, their provisions are included in these 1999 regs.

(2) The instruments specified in column 1 of Schedule 2 shall be amended in accordance with the corresponding provisions in column 3 of that Schedule.

(2) Schedule 2 contains a list of amendments of other regulations effected by these regulations.

Transitional provision

30. The substitution of provisions in these Regulations for provisions of the Management of Health and Safety at Work Regulations 1992 shall not affect the continuity of the law; and accordingly anything done under or for the purposes of such provision of the 1992 Regulations shall have effect as if done under or for the purposes of any corresponding provision of these Regulations.

Acts done under MHSWR 1992 continue to be legal as if done under equivalent provisions of MHSWR 1999

SCHEDULE 1
GENERAL PRINCIPLES OF PREVENTION
(This Schedule specifies the general principles of prevention set out in Article 6(2) of Council Directive 89/391/EEC) (OJ No. L183, 29.6.89, p.1)

(a) avoiding risks;

(b) evaluating the risks which cannot be avoided;

(c) combating the risks at source;

(d) adapting the work to the individual, especially as regards the design of workplaces, the choice of work equipment and the choice of working and production methods, with a view, in particular, to alleviating monotonous work and work at a predetermined work-rate and to reducing their effect on health;

(e) adapting to technical progress;

(f) replacing the dangerous by the non-dangerous or the less dangerous;

(g) developing a coherent overall prevention policy which covers technology, organisation of work, working conditions, social relationships and the influence of factors relating to the working environment;

(h) giving collective protective measures priority over individual protective measures; and

(i) giving appropriate instructions to employees.

See explanation at reg. 4 above

SCHEDULE 2
CONSEQUENTIAL AMENDMENTS

Column 1 Description of Instrument	Column 2 References	Column 3 Extent of Modification
The Safety Representatives and Safety Committees Regulations 1977	S.I. 1977/500; amended by S.I. 1992/2051; S.I. 1996/1513; S.I. 1997/1840; S.I. 1999/860 and by section 1(1) and (2) of the Employment Rights (Dispute Resolution) Act 1998.	In regulation 4A(1)(b) for "regulations 6(1) and 7(1)(b) of the Management of Health and Safety at Work Regulations 1992", there shall be substituted "regulations 7(1) and 8(1)(b) of the Management of Health and Safety at Work Regulations1999;".
The Offshore Installations (Safety Representatives and Safety Committees) Regulations 1989	S.I. 1989/971; amended by S.I. 1992/2885; S.I. 1993/1823; S.I. 1995/738; S.I. 1995/743; and S.I. 1995/3163.	In regulation 23(4) for "regulation 6(1) of the Management of Health and Safety at Work Regulations 1992", there shall be substituted "regulation 7(1) of the Management of Health and Safety at Work Regulations 1999".
The Railways (Safety Case) Regulations 1994	S.I. 1994/237; amended by S.I. 1996/1592.	In paragraph 6 of Schedule 1 for "regulation 3 of the Management of Health and Safety at Work Regulations 1992 and particulars of the arrangements he has made pursuant to regulation 4(1) thereof.", there shall be substituted "regulation 3 of the Management of Health and Safety at Work Regulations 1999 and particulars of the arrangements he has made in accordance with regulation 5(1) thereof.".
The Suspension from Work (on Maternity Grounds) Order 1994*	S.I. 1994/2930.	In article 1(2)(b) for ""the 1992 Regulations" means the Management of Health and Safety at Work Regulations 1992", there shall be substituted, ""the 1999 Regulations" means the Management of Health and Safety at Work Regulations 1999"; and In article 2(b) for "regulation 13B of the 1992 regulations", there shall be substituted "regulation 17 of the 1999 Regulations".

Column 1 Description of Instrument	Column 2 References	Column 3 Extent of Modification
The Construction (Design and Management) Regulations 1994	S.I. 1994/3140; amended by S.I. 1996/1592.	In regulation 16(1)(a) for "regulation 9 of the Management of Health and Safety at Work Regulations 1992", there shall be substituted "regulation 11 of the Management of Health and Safety at Work Regulations 1999;";
		In regulation 17(2)(a) for "regulation 8 of the Management of Health and Safety at Work Regulations 1992;", there shall be substituted "regulation 10 of the Management of Health and Safety at Work Regulations 1999;";
		In regulation 17(2)(b) for "regulation 11(2)(b) of the Management of Health and Safety at Work Regulations 1992", there shall be substituted "regulation 13(2)(b) of the Management of Health and Safety at Work Regulations 1999"; and
		In regulation 19(1)(b) for "the Management of Health and Safety at Work Regulations 1992", there shall be substituted "the Management of Health and Safety at Work Regulations 1999".
The Escape and Rescue from Mines Regulations 1995	S.I. 1995/2870.	In regulation 2(1) for ""the 1992 Regulations" means the Management of Health and Safety at Work Regulations 1992", there shall be substituted ""the 1999 Regulations" means the Management of Health and Safety at Work Regulations 1999"; and
		In regulation 4(2) for "regulation 3 of the 1992 Regulations," there shall be substituted "regulation 3 of the 1999 Regulations.".
The Mines Miscellaneous Health and Safety Provisions Regulations 1995	S.I. 1995/2005.	In regulation 2(1) for ""the 1992 regulations" means the Management of Health and Safety

Column 1 Description of Instrument	Column 2 References	Column 3 Extent of Modification
		at Work Regulations 1992;", there shall be substituted ""the 1999 Regulations" means the Management of Health and Safety at Work Regulations 1999;"; and
		In regulation 4(1)(a) for "regulation 3 of the 1992 Regulations;", there shall be substituted "regulation 3 of the 1999 Regulations;".
The Quarries Miscellaneous Health and Safety Provisions Regulations 1995	S.I. 1995/2036.	In regulation 2(1) for ""the 1992 Regulations" means the Management of Health and Safety at Work Regulations 1992;", there shall be substituted ""the 1999 Regulations" means the Management of Health and Safety at Work Regulations 1999;" and
		In regulation 4(1)(a) for "regulation 3 of the 1992 Regulations;", there shall be substituted "regulation 3 of the 1999 Regulations;".
The Borehole Sites and Operations Regulations 1995	S.I. 1995/2038.	In regulation 7(5) for ""the Management Regulations" means the Management of Health and Safety at Work Regulations 1992", there shall be substituted ""the Management Regulations" means the Management of Health and Safety at Work Regulations 1999.".
The Gas Safety (Management) Regulations 1996	S.I. 1996/551.	In paragraph 5 of Schedule 1 for "regulation 3 of the Management of Health and Safety at Work Regulations 1992, and particulars of the arrangements he has made in accordance with regulation 4(1) thereof.", there shall be substituted "regulation 3 of the Management of Health and Safety at Work Regulations 1999, and particulars of the arrangements he has made in accordance with regulation 5(1) thereof.".

Column 1 Description of Instrument	Column 2 References	Column 3 Extent of Modification
The Health and Safety (Safety Signs and Signals) Regulations 1996	S.I. 1996/341.	In regulation 4(1) for "paragraph (1) of regulation 3 of the Management of Health and Safety at Work Regulations 1992", there shall be substituted "paragraph (1) of regulation 3 of the Management of Health and Safety at Work Regulations 1999".
The Health and Safety (Consultation with Employees) Regulations 1996	S.I. 1996/1513.	In regulation 3(b) for "regulations 6(1) and 7(1)(b) of the Management of Health and Safety at Work Regulations 1992", there shall be substituted "regulations 7(1) and 8(1)(b) of the Management of Health and Safety at Work Regulations 1999".
The Fire Precautions (Workplace) Regulations 1997	S.I. 1997/1840; amended by S.I. 1999/1877.	In regulation 2(1) for ""the 1992 Management Regulations" means the Management of Health and Safety at Work Regulations 1992", there shall be substituted ""the 1999 Management Regulations" means the Management of Health and Safety at Work Regulations 1999"; In regulation 2(1) in the definitions of "employee" and "employer" for "1992" substitute "1999"; and In regulation 9(2)(b) for the words "regulations 1 to 4, 6 to 10 and 11(2) and (3) of the 1992 Management Regulations (as amended by Part III of these Regulations)", there shall be substituted "regulations 1 to 5, 7 to 12 and 13(2) and (3) of the 1999 Management Regulations".
The Control of Lead at Work Regulations 1998	S.I. 1998/543.	In regulation 5 for "regulation 3 of the Management of Health and Safety at Work Regulations 1992", there shall be substituted "regulation 3 of the Management of Health and Safety at Work Regulations 1999".

Column 1 Description of Instrument	Column 2 References	Column 3 Extent of Modification
The Working Time Regulations 1998*	S.I. 1998/1833.	In regulation 6(8)(b) for "regulation 3 of the Management of Health and Safety at Work Regulations 1992", there shall be substituted "regulation 3 of the Management of Health and Safety at Work Regulations 1999".
The Quarries Regulations 1999	S.I. 1999/2024.	In regulation 2(1) for ""the 1992 Regulations means the Management of Health and Safety at Work Regulations 1992;", there shall be substituted ""the 1999 Regulations" means the Management of Health and Safety at Work Regulations 1999;". In regulation 7(1)(a) for "paragraphs (1) to (3c) of regulation 3 of the 1992 Regulations;" there shall be substituted "regulation 3 of the Management of Health and Safety at Work Regulations 1999". In regulation 43 for "regulation 5 of the 1992 regulations" there shall be substituted "regulation 6 of the 1999 Regulations".

Note

The Regulations marked with an asterisk are referred to in the Preamble to these Regulations.

Schedule 2

See explanation at reg. 29 above.

OLA	Occupiers Liability Acts, 1957 & 1984
OSRPA	Offices Shops and Railways Premises Act, 1963 (now largely repealed)
PGCE	Post Graduate Certificate in Education.
PPER	Personal Protective Equipment at Work Regulations, 1992
PUWER	Provision and Use of Workplace Equipment Regulations, 1998
QUANGO	Quasi Autonomous non governmental organisation.
RoSPA	Royal Society for the Prevention of Accidents
reg.	Regulation.
RSP	Registered Safety Practitioner (IOSH)
s	section (of a Statute)
SEA	Scottish Environment Agency
SI	Statutory instrument/delegated legislation/regulations, etc.
SoS	Secretary of State for Education and employment
SRSC	Safety Representatives and Safety Committee Regulations, 1977
TU	Trade Union
TUC	Trades Union Congress
UK	United Kingdom of Great Britain and Northern Ireland.
WPHSWR	Workplace (Health Safety and Welfare) Regulations, 1992
WTR	Working Time Regulations, 1998

Index